The Knowledge Medium: Designing Effective Computer-Based Learning Environments

Gary A. Berg
California State University, Channel Islands, USA

 **Information Science Publishing**
Hershey • London • Melbourne • Singapore • Beijing

Acquisitions Editor:	Mehdi Khosrow-Pour
Managing Editor:	Jan Travers
Development Editor:	Michele Rossi
Copy Editor:	Jane Conley
Typesetter:	Amanda Appicello
Cover Design:	Integrated Book Technology
Printed at:	Integrated Book Technology

Published in the United States of America by
    Information Science Publishing (an imprint of Idea Group Inc.)
    701 E. Chocolate Avenue
    Hershey PA 17033-1117
    Tel: 717-533-8845
    Fax: 717-533-8661
    E-mail: cust@idea-group.com
    Web site: http://www.idea-group.com

and in the United Kingdom by
    Information Science Publishing (an imprint of Idea Group Inc.)
    3 Henrietta Street
    Covent Garden
    London WC2E 8LU
    Tel: 44 20 7240 0856
    Fax: 44 20 7379 3313
    Web site: http://www.eurospan.co.uk

Copyright © 2003 by Idea Group Inc. All rights reserved. No part of this book may be reproduced in any form or by any means, electronic or mechanical, including photocopying, without written permission from the publisher.

Library of Congress Cataloging-in-Publication Data

Berg, Gary A., 1955-
  The knowledge medium : designing effective computer-based learning environments /
  Gary A. Berg.
    p. cm.
  Includes bibliographical references and index.
  ISBN 1-59140-103-8 (cloth)
   1. Computer-assisted instruction. I. Title.

LB 1028.5 .B4362 2002
371.33'4--dc21   2002068782

eISBN 1-59140-111-9

British Cataloguing in Publication Data
A Cataloguing in Publication record for this book is available from the British Library.

NEW Titles from Information Science Publishing

- **Web-Based Education: Learning from Experience**
  Anil Aggarwal
  ISBN: 1-59140-102-X; eISBN 1-59140-110-0, © 2003

- **The Knowledge Medium: Designing Effective Computer-Based Educational Learning Environments**
  Gary A. Berg
  ISBN: 1-59140-103-8; eISBN 1-59140-111-9, © 2003

- **Sociotechnical and Human Cognition Elements of Information Systems**
  Steve Clarke, Elayne Coakes, M. Gordon Hunter and Andrew Wenn
  ISBN: 1-59140-104-6; eISBN 1-59140-112-7, © 2003

- **Usability Evaluation of Online Learning Programs**
  Claude Ghaoui
  ISBN: 1-59140-105-4; eISBN 1-59140-113-5, © 2003

- **Building a Virtual Library**
  Ardis Hanson & Bruce Lubotsky
  ISBN: 1-59140-106-2; eISBN 1-59140-114-3, © 2003

- **Design and Implementation of Web-Enabled Teaching Tools**
  Mary F. Hricko
  ISBN: 1-59140-107-0; eISBN 1-59140-115-1, © 2003

- **Designing Campus Portals**
  Ali Jafari and Mark Sheehan
  ISBN: 1-59140-108-9; eISBN 1-59140-116-X, © 2003

- **Challenges of Teaching with Technology Across the Curriculum: Issues and Solutions**
  Lawrence A. Tomei
  ISBN: 1-59140-109-7; eISBN 1-59140-117-8, © 2003

Excellent additions to your institution's library! Recommend these titles to your Librarian!

To receive a copy of the Idea Group Inc. catalog, please contact (toll free) 1/800-345-4332, fax 1/717-533-8661, or visit the IGP Online Bookstore at: http://www.idea-group.com!

Note: All IGI books are also available as ebooks on netlibrary.com as well as other ebook sources. Contact Ms. Carrie Stull at <cstull@idea-group.com> to receive a complete list of sources where you can obtain ebook information or IGP titles.

Dedication

This book is dedicated to my father. Although he passed before finishing his design book, he gave me an appreciation for the beauty of ideas, and a belief that what I think is worth writing down (even when I barely made sense). Dad, I hope I make more sense now.

The Knowledge Medium: Designing Effective Computer-Based Learning Environments

Table of Contents

Preface i
    Gary A. Berg, California State University, Channel Islands, USA

Chapter I. Introduction 1

PART I:
Computer-Based Learning Theory and Practice

Chapter II. Learning Theory and Technology 9
    Educational and Instructional Film 10
    Audio Recordings and Radio for Educational Purposes 12
    Educational and Instructional Television 12
    Programmed Instruction and Computer-Assisted Instruction
        Movements 14
    Learning Theory: Behaviorism Versus Constructivism 14
    Piaget 15
    Vygotsky and Cooperative Learning Theory 16
    Skinner 17
    Learning Styles, Multiple Intelligences, and Self-Regulated
        Learning 18
    Adult Learning Theory 19
    Conclusion 26

Chapter III. Tutorial Method: Profiling, Customization and
    Agents 28
    Specific Techniques 34
    Conclusion 38

Chapter IV. Group Method: Virtual Teams and Communities 39
    To Group or Not to Group 45
    Traditional Notions of Community in Higher Education 46
    Social or Group Learning Theories 47
    Research Literature on Group Method in Computer
        Environments 49
    Virtual Teams in Business 51
    Interdependent Tasks 52
    Leadership 52
    Social Communication 53
    Project Management 53
    Threaded Conversations/Chat 53
    Computer-Mediated Conferencing 54
    Conclusion 55

Chapter V. Human-Computer Interaction in Education 60
    Human Factors 61
    Usability 62
    Interface Design 63
    GOMS Models 63
    Command Language Versus Direct Manipulation 64
    Hypertext 65
    Graphic and Visual Issues 66
    Metaphor 66
    Animation 67
    Organizational Issues 68
    Task Analysis 68
    Conclusion 69

Chapter VI. Interactivity and Navigation 70
    Interaction with Media 78
    Conclusion 84

Chapter VII. Computer Tools for Learning 86
    Critical Thinking 86
    Concept Mapping 87
    Teaching with Programming 90
    Mindtools - Cognitive Amplification 91
    Conclusion 92

Chapter VIII. Faculty and Teaching Issues 93

PART II:
Computer as Medium

Chapter IX. Film Theory 104
    Current Practices 105
    History of Film 110
    Ideology and Film 111
    Academic Literature on Film and Education 112
    Computers as Medium 114
    Communications and Media Theory 115
    Still Photography 118
    Phenomenology of Film 121
    I Am Lying: Documentary Film and Truth 122
    Alternative Strategies 125
    Knowledge Media and Mixing Media 127
    Recent Film Theory: Semiology 128
    Cognitive Film Theory 131
    Conclusion 132

Chapter X. Dramatic Structure, Genre, and Editing 134
    Use of Dramatic Structure 135
    Narrative Conventions: Premise and Time 136
    Genre and Audience Expectations 139
    Chance and Randomness 141
    Editing 141
    Editing Strategies 142
    Eisenstein 144
    Narrative Versus Non-Narrative 150
    Sound 155
    Conclusion 157

Chapter XI. Subjectivity, Point of View, and Dreaming 159
    Point of View and Subjectivity 159
    Media Viewing and Dreaming 163
    The Wish 168
    The Hero 169
    Bunuel 170
    Conclusion 174

Chapter XII. Stories, Simulations, and Case Studies 175
    Stories to Organize Learning Environments 179
    Review of Psychology and Education Literature 179
    Narrative Structure in Hypertext Environments 183
    Case-Based Learning/Simulations 184
    Conclusion 186

Chapter XIII. Conclusion – How We Might Learn 187
    Current Practices 190
    What We Know 191
    Visions for the Future 199
    Learning Devices 202
    Research Agenda 205
    Final Thoughts 205

References 207

Appendices 221
    Appendix A: Research Methodology 221
    Appendix B: 2000 Survey Quantitative Instrument 224
    Appendix C: 2000 Qualitative Instrument 240
    Appendix D: 2000 Survey Cover Email 243
    Appendix E: 2000 Survey Email to Interview Subjects 244
    Appendix F: 2000 Survey Online Statement of Informed
        Consent 245
    Appendix G: 2000 Survey In-Person Statement of Informed
        Consent 246
    Appendix H: 2000 Survey Post-Interview Form 248
    Appendix I: 2000 Survey List of Participating Institutions 249
    Appendix J: 2001 Survey Letter to Subjects 256
    Appendix K: 2001 Quantitative Instrument 257

About the Author 260

Index 261

Preface

I worked for three years putting together training programs for film and television directors while employed at the Directors Guild of America. At that time, the early 1990s, directors were clearly interested in learning all they could about the use of computers both for evolving traditional video and film forms and what was then called "multi-media." The use of computers in digital editing and special effects was discussed widely, but additionally many directors were aware that a new medium was emerging, one that may be a source of future employment for them. Consequently, they flocked to the educational programs I developed—both lectures on advanced uses of computers in film and video production and hands-on laboratories where they learned tools for creating new media. I also arranged for demonstrations and tours of multimedia production facilities. Ill-fated early companies such as Phillips CD-I, 3DO, and Answer Interactive Television are some of the companies that come to mind. It was fascinating to observe Hollywood directors as they watched these demonstrations because they understood the way new computer applications resembled the film and video production process they were expert in, but also how they were different. I could tell by their questions, particularly about user interaction issues, that they understood that this was a new medium.

Almost a decade has passed since this experience, but the impression stayed with me. As I began to work as an administrator in broader educational subject matter and became involved in developing CD- and web-based courses, I remembered the reaction of the film and television directors. They were searching for an understanding of the principles of this new medium, and how they as "authors" could use this new technology to their advantage. For this group of Hollywood directors the biggest issue was control: they are accustomed to controlling viewers in very precise ways. In fact, one way great Hollywood filmmaking is defined is by the degree to which the director is in control of the audience's point of view and effectively manipulates its members into thinking and feeling specific ways at particular times. Alfred Hitchcock, the great director, is a perfect example of the type of control directors often exercise. He is said to have so meticulously engineered his movies in advance that he admitted to boredom during the actual shooting of his movies. Many contemporary directors such as Spielberg, Lucas, and Coppola exert similar control over the viewing experience. While some other independent-style directors such as Mike Leigh, Woody Allen, and others have more improvisation styles, they generally are giving up control to the actors, not the audience. It seems to me that this director control of film and television is not unlike the control faculty members have in the traditional classroom. The most dominant form of educational delivery in higher education is still the lecture, and generally effective teachers are identified by their ability to present and control the classroom. While certainly one can be an active listener in a lecture, or actively view a film, that sort of activity is nothing like the potential control learners can gain in interactive new media.

I still remember vividly the hush that went over the audience of film and television directors when an executive from an interactive television company told of plans to give viewers the ability to choose camera angles and lens length during televised football games. Of course, on the surface they were concerned about their jobs, but even more importantly, when viewers have this kind of control, it is a different medium. The horror these directors experienced at the thought of giving up control to viewers is not unlike that expressed today by some faculty members at the notion of students controlling their own educational experience.

I come to the subject of this book from a very different path than most of those thinking about the use of computers in educational environments. My formal education focused originally on literature and film studies, and film production at the University of California at Berkeley, San Francisco State University, and the University of California at Los Angeles. I became professionally involved in educational administration through the backdoor of continuing education focused first on the entertainment industry, and then more broadly. It was after this combined experience of studying film and television and working in adult education that I began research in education and earned a doctorate in the field of higher education from Claremont Graduate University, with a special emphasis on distance learning. I hope that the different point of view I have developed from my eclectic background gives me the ability to make something of a unique contribution to this evolving new field.

Acknowledgments

Many of the ideas presented in this book have appeared in early form in various publications. I'd like to acknowledge the following publications for their initial support of my scholarship in this field: *WebNet Journal: Internet Technologies, Applications & Issues*, *Journal of Educational Multimedia and Hypermedia*, *Education at a Distance Journal*, *Journal of Asynchronous Learning Networks*, and *Education Policy Analysis Archives*. Additionally, I'd like to thank Magid Igbaria at Claremont Graduate University for his initial encouragement and inspiration in the study of virtual communities and the social aspects of the use of computers. Finally, I want to acknowledge the contribution of my wife, Linda Venis, for her careful editing and thoughtful suggestions.

Chapter I

Introduction

Designing for the little screen on the desktop has the most in common with designing for the Big Screen. Interactive software needs the talents of a Disney, a Griffith, a Welles, a Hitchcock, a Capra.... (Nelson, 1995, p. 243)

When sound finally came to film in 1927 with *The Jazz Singer*, producers went first to the Broadway stage looking for actors, writers, and directors. The first projects they developed were often little more than filmed stageplays. In some ways the technology of the time dictated this strategy—the camera had to be enclosed in a large box in order to deaden the sound of the celluloid strips moving through the machine, and the microphones had limited range, forcing the actors to stay close when speaking. More importantly, there was a conceptual limitation. Filmmakers had not yet developed a unique language for sound cinema, and they therefore began by mimicking theater. We are now in a very similar stage in the development of technology-enabled education. Distance-learning courses offered by videotape, online, and computer still use the language of the conventional classroom. For the most part, these courses with their talking heads and online campus metaphors do little more than automate the traditional classroom. Computer-based learning has yet to find its own language and is still metaphorically locking the camera to the floor and recording the traditional classroom, rather than offering a new approach to learning.

Figure 1: Sound-Proof Booth (Library of Congress, Prints and Photographs Division [LC-US262-95604])

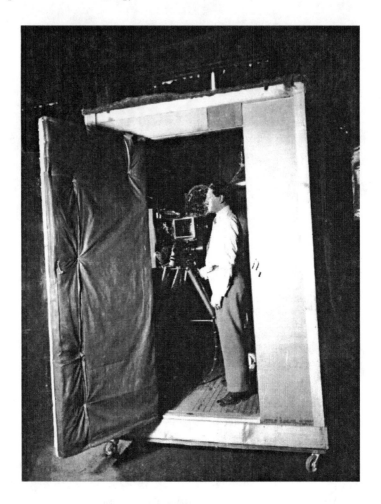

Alan Kay (1995) quotes Marshal McLuhan in suggesting that if the personal computer is a truly new medium then the very use of it will change cultural and individual thought patterns. Could the development of this new medium change education? Kay argues that in order for users to receive messages embedded in a medium, they need to have internalized the medium. While the American film has developed an elaborate symbolic sound and image code over the years—the conventions clearly understood by the general viewing audience—computers have yet to develop such complex viewing conventions. We need to develop these computer-viewing conventions particularly for learners.

Although there are similarities between computers and other media (film in particular), there are unique properties as well that are transformative to the user. Printed books transformed society by allowing users to preserve and share information. The computer and digital communications are again transforming society in a similarly large way. The important questions then are: What are the specific properties of the new medium? What content is best delivered through this medium? Westland asks, "If the medium is the message, then what kinds of messages are facilitated by multimedia?" (Westland, 1994, p. 359).

Furthermore, if computers are a medium, which other media are most similar? Nelson (1995) argues that movies have the greatest experience in working with psychological and visual effects on screens and for this reason film is most relevant as a medium to computers. He points out that particular abilities are required for the effective use of both computers and film as a medium—most importantly, a unifying vision. Baecker and Small (1995) agree with Nelson in arguing that designers should look to the language of cinema for models of how computer interfaces are structured. Cinematic theory and conventions developed from American film integrated with new early computer conventions might lead to the establishment of a computer medium language. This computer medium language is one that is essential for the further development of educational software.

The history of educational technology shows a pattern of moments of exaggerated promise at the introduction of new technology, followed by disappointment. Thomas Edison predicted in 1913 that books would be replaced by motion pictures (Cuban, 1986). In 1940, George F. Zook in his American Council on Education report described film as "the most revolutionary instrument introduced in education since the printing press" (Hoban, 1942, p. 16). However, after these early periods of great promise, the history of the use of technology in education is one of resistance to change and disappointment. While film came into wide use in educational environments during Word War II when the military needed a device to speed up the training of masses of soldiers with various skill levels and education, it never gained acceptance in higher education in the same way. Hoban blames resistance to educational technology from educators partially on the Puritanical belief in the power of words, and a suspicion of any education that seems pleasurable. Whatever the reasons, the lack of success is clear.

The research literature on the use of film and TV in educational environments is striking in that one finds much written and published in the period of the 1930s-50s, and then very little afterwards. Research in the uses of film in education had, in the opinion of one of the leading researchers in this area, remained almost at a standstill between the 1950s and 1970s (Hoban, 1971). In the late sixties and seventies, a few authors concentrated on how to use films to teach creatively as an augmentation and resource in the classroom (Schillaci & Culkin, 1970; Worth,

1981), while others argued about the educational value of film and television, especially Sesame Street (Goldman & Burnett, 1971; Cook, et al., 1975). Overall, there is surprisingly little written about the use of film and television in education.

With the introduction of the personal computer, large claims were once again made for educational applications. The programmed learning, or auto-instructional movement began with the introduction of computers, and early on emphasized B.F. Skinner's model of operant conditioning, response mode, error rate, and reinforcement (DeCecco, 1964). Later, Computer-Aided Instruction (CAI), and then Intelligent Computer-Aided Instruction (ICAI) developed, seeking to combine artificial intelligence capabilities (Frasson & Gauthier, 1990). However, neither of these movements had much success in education at any level.

Educational theorists have praised technology and in many cases have also exaggerated its promise. For constructivists, computers may finally provide the means by which the very labor-intensive educational philosophy of Dewey may be put into practice. On the other side of the fence, behaviorists have long held sway in the field of computer-based training (CBT) with the tireless repetition and utilization of clear behavioral learning objectives–key elements of these training programs.

When and how will technology-enabled learning find its own mature language? Film came to its own language through the artful incorporation of techniques from many different fields including fine art, philosophy, drama, literature, and newspaper comics of the late 19th century. The same kind of coming together of ideas and techniques from multiple fields including psychology, computer programming, graphic arts, and film theory is going to result in the new language of technology-enabled learning. As with the film industry, the convergence of these different disciplines will result eventually in a new language, coming of age in the real world experience of learners using these educational approaches in a successful manner.

This book is about that new language. What follows is an attempt to spark a discussion that will lead to answers to the question of what are the most effective techniques for the design of computer learning environments. This is not a how-to book—we are too early in the evolutionary process of the medium to give such specific guidance. Rather, my intention is to offer some theories to elevate the thinking bout computers in education. Because the subject is interdisciplinary, combining science with the humanities, the theoretical discussion draws from a broad range of disciplines: psychology, educational theory, film criticism, and computer science. The book looks at the notion of computer as medium and what such an idea might mean for education. I suggest that the understanding of computers as a medium may be a key to re-envisioning educational technology. Oren (1995) argues that understanding computers as a medium means enlarging human-computer interaction (HCI) research to include issues such as the psychology of media, evolution of genre and form, and the societal implications of media, all of which are discussed here. Computers began to be used in educational environments much later than film, and I would have to agree with others who claim that the use of computers instructionally is still quite unsophisticated.

This book is divided into two parts: the first part focuses on computer-based learning theory and practice, while the second part looks at the issues connected with considering the computer as a medium. The first half of the book is descriptive; the second half is theoretical. In the first chapter, a brief analysis of the history of both educational technology and relevant learning theory is sketched out. In the next two chapters, the two main approaches (tutorial and group) to distance learning are examined. In Chapter Two, the specific methods and issues connected to the tutorial methods are discussed with a particular emphasis on customization, profiling, and agents. In the third chapter, group learning methods are investigated. The research literature on virtual teams, interdependent tasks, community formation, and specific techniques for organizing group tasks in computer educational environments are discussed. Then in Chapter Four the literature on human-computer interaction is examined with an eye towards understanding how it may be especially relevant to education. What are the current issues in HCI research? What aspects of HCI research are relevant to education? In the following chapter, the all-important issues of interactivity and navigation in educational environments are then examined. What is the nature of interactivity in educational software? What are the challenges for users navigating through hypertext environments? In the next chapter, computer tools for learning such as concept maps, knowledge modeling, and learning through computer programming are considered. The first part of the book ends with a look at teaching issues uncovered in the results reported from two new surveys.

In Part Two of the book, I turn to considering the ramifications of understanding computer software used for educational purposes as a new medium. Traditionally, film critics have been more concerned with arguments about authorship and highly political and jargonistic debates about feminism, psychoanalysis, and semiology as applied to film. There has been very little written by film critics on educational media. Using the background on educational technology, learning theory, and specific computer-based learning issues, I look at designing educational software through the lens of film criticism. In Chapter Eight, broad issues such as phenomenology and ideology of film are examined, as well as the history of moving and still photography, documentary film and recent film theory. In the following chapter, dramatic structure is considered in terms of specific issues such as genre, conventions and editing. In Chapter Ten, psychoanalytic topics including subjectivity, point of view, and dreaming are addressed. In Chapter Eleven, the ramifications of the recent narrative psychology movement on computer environments is examined, particularly as they relate to the use of case studies and simulations.

Additionally, the data collected from two original studies from both the administrative and student perspectives on distance learning pedagogical practices in higher education are reviewed in the book and connected to the theoretical information presented. The 2000 survey focuses on the administrative perspective in interviews with representatives from 17 institutions, and quantitative data from 176 universities. The 2001 survey is based on responses from 129 students who have completed distance learning courses in various formats centering on their attitudes and experiences in these courses. The increased use of distance learning is one of the most controversial trends in American higher education today. Nevertheless, clear and comprehensive information on how distance learning is being implemented in higher education is hard to find. Previous studies have focused primarily on quantifying the rate of increased use and attempts to catalogue brief descriptions of the programs available at various institutions. What have not been addressed adequately in previous research are the questions of pedagogical practices as they relate to using educational technology. With the help of the data from these two surveys, I believe readers will have a better understanding of pedagogical practices in distance learning from both the administrative and student perspectives.

Those who run distance learning programs in American universities are very optimistic about their use. One of the studies (2000) conducted for this book found that most administrators (69.4%) responsible for distance learning believed that the delivery format represents a pedagogical innovation.

Figure 2: Distance Learning as a Teaching/Learning Innovation (Question 38)

Do you view distance learning in higher education as a teaching/learning innovation?

		Frequency	Percent	Valid Percent	Cumulative Percent
Valid	yes	118	67.0	69.4	69.4
	maybe	41	23.3	24.1	93.5
	no	11	6.3	6.5	100.0
	Total	170	96.6	100.0	
Missing	System	6	3.4		
Total		176	100.0		

Why are they so optimistic? I examine in this book both how administrators are designing computer-based courses, and how students are responding.

Finally, the book ends with a summary of current practices as revealed through the collected data and recommendations for best practices based on what we now know from the research discussed in the book. I then look to the future and offer a vision for how we might learn. In this way, I hope to contribute to the field and in so doing help to make the clear optimism of the university administrators justified.

Part I

Computer-Based Learning Theory and Practice

Chapter II

Learning Theory and Technology: Behavioral, Constructivist, and Adult Learning Approaches

Educational technology is hampered by the absence of a theory of the structure of symbols and their effects or function relating to the mediation of cognitive processes. For example, it is still not clear how the structure of information in film differs from that in pictorial representation or in language (Saettler, 1990, p. 263).

This chapter briefly reviews the history of educational technology, surveys learning theory, and sketches out directions that might contribute to the needed theory Saettler points to above. In order to understand fully educational software, one needs to have a sense of the history of educational technology. Learning theories that have been applied to computer environments also need close examination because they provide the rationale for specific approaches and learning strategies. In particular, we will see in this chapter the important influence behavioral and constructivist theoretical models have had on software design. Additionally, adult learning theories are investigated because of their relevance to educational software in post-secondary education. Finally, cooperative learning and learning style theories are outlined because of their relevance to specific computer-based teaching methods.

According to Paul Saettler (1990), one of the leading historians in the field, educational technology in a broad sense can be traced back to when tribal priests presented systematized knowledge and used early sign writing to pass on knowledge. He points out that the more advanced the culture, the more complex educational technology became. Saettler observes that there is wide disagreement concerning theories in the field of educational technology. Furthermore, he argues that learning theory is a key to the field because behavioral sciences applied to problems of learning and instruction are fundamental to an evolving educational technology. Saettler traces the beginning of educational technology in America to the use of lantern slides for educational purposes in the American Lyceum and Chautauqua movements. Particularly in the Chautauqua educational programs aimed at adults wanting to continue their education, the slides were an integral part of the way programs were presented to large audiences.

Later, what became known as the Visual Instruction Movement promoted visual approaches to teaching to combat what they labeled "verbalism" in the classroom. One influential figure in this movement was Edgar Dale, who developed a schema he called the "Cone of Experience" emphasizing experiential learning over verbal symbols. The Visual Instruction Movement was concerned primarily with the use of specific media, while educational technology is oriented more towards psychological principles and the total teaching-learning process.

In addition to the Visual Instruction Movement, Saettler identifies the Lancasterian method of instruction involving systematic instruction around memorization and drill as an early influence on educational technology, particularly behaviorist approaches. Behaviorism emphasized behavior of the learner and reinforcement, and was first introduced by John B. Watson in a 1913 article, and then developed most extensively later by B.F. Skinner. Skinner saw the curriculum as forming behavioral objectives, or the arrangement of contingencies of reinforcement. In relationship to the progressive and constructivist learning theory, Saettler identifies Thorndike as the first modern instructional technologist, replaced later only by John Dewey in influence. The Dewey-inspired Progressive Movement provided the philosophical grounding for what later became constructivism. So we can see that early on there were two main techniques for the use of media in the classroom: one concentrating on behavior, and another focusing on process and wholistic, experiential-based learning.

EDUCATIONAL AND INSTRUCTIONAL FILM

A pioneer in film technology and an intellectual interested in education, Thomas Edison was a regular participant in Chautauqua Movement education programs and was one of the first to produce films for classroom use. In 1911 he released a series of historical films about the American Revolution, beginning with *The Minute Men*, followed by a series on natural and physical sciences in 1914. In addition to Edison, many colleges and universities were involved in educational film production as early as the 1910s with Yale University and the University of Minnesota being large players in this early film production. As early as 1910, the first education film catalog, the *Catalogue of Educational Motion Pictures*, listed 1,065 titles. Undoubtedly, many of these films were simple shorts documenting events, people, and places in an unsophisticated manner. Demand for training films during World War I led to an increase in educational film production. However, by the late 1920s and early 1930s, educational film companies had financial troubles and production slowed.

World War II was a turning point for educational film both in terms of number and technique. The use of education films was part of the official policy of the War Department, and consequently this led to the production of six times more educational films than had been created up to that point. Film production during the war led to two stylistic or pedagogical improvements, mostly brought about by the influx of Hollywood filmmakers into the military: first-person camera angles and camera movement. Overall, the influence of Hollywood filmmakers in the military led to the application of dramatic techniques used previously only in entertainment films. After World War II, additional universities became involved in educational filmmaking including the University of Chicago, University of Southern California, Ohio State, University of Wisconsin, New York University, the University of Indiana, University of Minnesota, Iowa State University, University of Michigan, Boston University, and Syracuse University.

One of the more interesting forms of post-World War II educational film was the development of the so-called "trigger" films. Orner (1990) defines trigger films as short, problem-centered films designed to engage students in discussion. Trigger films are significant in educational technology because they are an example of an attempt to stimulate viewers through popular film techniques of subjective viewer positioning. We will see later in Chapter Ten, in the discussion on point of view and subjectivity, how important this issue is in constructing computer-based educational environments. One ideological problem with these films is that they ignore the socio-cultural differences of students by using subjective camera angles to involve the viewers. Also, the lack of narrative closure in these educational films gives the illusion of openness to alternative points of view, when this is not the case.

The first academic research on instructional films was done in 1912, and the first large-scale research was done by Johns Hopkins University in 1919 analyzing the effectiveness of venereal disease training films. We will see in Part Two of this book that the research on educational use of instructional film is remarkably thin. Nevertheless, one can see from this brief history that the power of film used for education and training was recognized by many, and that often fiction film strategies were used.

AUDIO RECORDINGS AND RADIO FOR EDUCATIONAL PURPOSES

Thomas Edison's invention of the tin foil phonograph in 1877 made the first language laboratories possible, leading to its first use in 1891 in a foreign language class offered at the College of Milwaukee. In 1900, a professor at the University of California taught Chinese concurrently at the University of Pennsylvania using a wax cylinder sent across the United States. Thomas J. Foster's International Correspondence Schools used the audio cylinders for foreign language courses early on as well. The methods and procedures for language correspondence courses were developed in this early period and applied later using more sophisticated technologies including video and film (Kitao, 1995).

After World War I, university-owned radio stations became a common phenomenon. Educational radio grew greatly from 1925 to 1935, and by 1936 there were 202 such stations across the country. At its height, thirty-six institutions offered courses through combining radio and mail delivery and thirteen universities offered the courses for credit. "Schools of the Air" were founded at the University of Wisconsin, University of Kansas, University of Michigan, and the University of Minnesota. At first the courses simply involved professors reading lectures over the radio, but they grew in sophistication and production quality. By the end of the World War II, most of the educational radio efforts were discontinued. Federal regulation, the rise of commercial radio networks, limitations of the broadcast signal, lack of a target population, and minimal faculty involvement were reasons identified for the failure (Saettler, 1990; Pittman, 1986).

EDUCATIONAL AND INSTRUCTIONAL TELEVISION

In the research literature, "educational television" usually refers to programs that have a broad cultural purpose such as *Sesame Street*, while "instructional television (ITV)" is generally used to describe videotaped whole courses. The first instructional television station was started in 1953 by the University of Iowa, and by the 1960s, they were used widely by other institutions in the United States. The breakthrough in instructional television came with the use of trained staff, faculty used on a rotating basis, improved studio resources, the development of a nationwide network, and an increased commitment by educators to use television. The 1958 National Defense Education Act (NDEA), Title VII, specifically provided for presenting academic subject matter through media and thus encouraged its development. In the 1970s, community colleges developed telecourses to deliver coursework, which have found a steady audience up until the present time. Course developers such as the Public Broadcast System (PBS) (funded partly by the Annenberg Foundation), Dallas Telecourses, and Coast Community College regularly supply course content to other universities through licensing agreements for broadcast and cassette tape distribution.

De Vaney (1990) reported that the Ford Foundation played a key role in the development of instructional television, believing as early as 1972 that the solution to the problem of teacher shortage was instructional television. However, television eventually lost the interest of funders who instead turned in more recent years to the delivery of education through microcomputers. When broadcast radio stations were folding in the 1950s, staff members were often hired by new instructional television stations. This led to an emphasis on the audio track of the medium. Additionally, it should be pointed out that the history of commercial television is closer to radio than film, so in many ways instructional television is closer stylistically to radio than film. Furthermore, professors accustomed to the lecturing pushed ITV towards a radio format. Consequently, early ITV was jammed with facts delivered by lecture, with little use of visuals. ITV failed because of a resulting stilted format, limited student participation, lack of faculty trained for television, and curriculum that often did not take into account local area curriculum differences.

Kent and McNergney (1999) claim two factors led to the de-emphasis of instructional television: lack of quality programming, and the teacher-less approach to the medium. The teacher-less approach does not allow teachers to control or interact with the medium, and students are unable to raise questions or respond to media such as film, radio, and television. Kent and McNergney argue that low technologies such as chalkboard and textbooks have been more successful because they work in teacher-defined curriculum and are more flexible and durable. They see pedagogical flexibility, teacher control, and accessibility as the key issues in adoption of technology in classroom.

Research in the later half of the twentieth century showed no significant differences between instructional television and traditional classroom experiences. Recently, a book entitled *The No Significant Difference Phenomenon* (Russell, 1999) cataloged over 400 studies indicating equal learning outcomes. The NEA and AFT subsequently published a rebuttal claiming poor research methods in these 400 studies (The Institute for Higher Education Policy, 1999). Research on *Sesame Street* showed that educational effectiveness was limited by the absence of support materials and an interested adult to encourage and enrich lessons.

Nevertheless, instructional television was often seen as entertainment, and therefore not a serious part of the regular curriculum. In the 1980s, studies began on media comparisons concerned with developmental cognitive processes and their relationship to specific media attributes. Aptitude-treatment (ATI) research concentrated on which attributes were most effective. New theories of active TV viewing were brought forth, arguing against the previous notion of passive television viewing. Specific attributes of viewers in connection with TV were also investigated.

The history of educational and instructional television shows that their influence never met early expectations because of both administrative issues and limitations for the medium. Computers quickly began to steal the thunder of ITV with the promise of increased interactivity while still using video clips.

PROGRAMMED INSTRUCTION AND COMPUTER-ASSISTED INSTRUCTION MOVEMENTS

As early as the 1950s, educational technology began to incorporate the computer. The Programmed Instruction Movement revived individualizing instruction notions and although it declined, its influence lived on past the 1960s. Computer-Assisted Instruction (CAI) was first used in the 1950s, with much of the early work done at IBM. CAI growth occurred in the mid-1960s, but faded quickly by the late 1960s. The typical CAI program modes were drill-and-practice and tutorial with a strong degree of author or instructor, rather than learner control. Later, Intelligent Computer-Assisted Instruction (ICAI) or Intelligent Tutoring Systems (ITS) were developed from a cognitive science approach to educational technology. These computer-based instructional movements were influenced by specific learning theories. In order to understand these early computer-based education movements and the history of educational technology, one must have a basic knowledge of the learning theories behind the technologies.

LEARNING THEORY: BEHAVIORISM VERSUS CONSTRUCTIVISM

From behaviorists to constructivists, educational theorists have often praised technology, and exaggerated its promise. For constructivists, computers may finally provide the means by which the very labor-intensive educational philosophy of Dewey may be put into practice. In more recent times, those interested in concept mapping, the value of learning computer programming, and full-immersion simulations have all to some degree based their approaches on constructivism. When considering cognitive models, the task of educational technology is to focus on knowledge construction and to understand what cognitive structures users bring to the learning environment. On the other side of the fence, behaviorists have long held sway in the field of computer-based training (CBT) with the tireless repetition and utilization of clear behavioral learning objectives as a key element of these training programs. A brief summary of the behaviorist and the constructivist positions on technology in education and the key theorists is useful for charting future directions for technology-enabled education.

PIAGET

Jean Piaget, a primary influence on constructivist theory, developed a theory of learning connected to human development through his observation and study of children. Piaget's background was in science, and thus he was heavily influenced by notions of Darwinian evolution in understanding learning in children. After studying adaptation of organisms to their environment, Piaget reasoned that an organism's intelligence was embodied in structures with latent tendencies for development that could be brought out by appropriate interaction with the environment (Gardner, 1973). The research problem of intelligence for Piaget was to try to discover the different methods or tendencies for thinking used by children at various ages.

Piaget is known for discovering the notion of conservation in learning. He uses "equilibrium" to describe the adjustment between the individual's cognitive structures and his environment. Adaptation is considered in terms of two processes: "assimilation" and "accommodation." Assimilation involves the learner's interacting with the environment in relation to his own structures, while accommodation involves the transformation of his structures in response to the environment (Ginsburg & Opper, 1988). Assimilation and accommodation are fundamental in learning because they explain the process of integration of reality into a preexisting cognitive structure. For Piaget, learning occurs through successive stages of disequilibrium and equilibrium brought about by the effort to adapt to the real world (Campbell, 1976).

However, Piaget believed that the individual assists in the construction of his own reality. His notion of mental organization is that there is a tendency, both physical and psychological, to integrate into existing mental structures. In practical terms, Piaget promoted the notion of understanding as involving individual action. Consequently, teachers should set up situations that lead the child to question, to experiment, and to discover facts and relationships themselves, rather than through the teacher. Furthermore, in applying the notion of assimilation and accommodation, Piaget argued that learning is best facilitated if the experience presented to the child bears some relevance to what he already knows, while at the same time being sufficiently new to present conflicts. Piaget argued that traditional, large group learning should be avoided, and instead that children should work freely on individual projects (Ginsburg & Opper, 1988).

Piaget's theories have influenced educational computer software design. Ben Shneiderman (1993), a leading computer scientist specializing in computer interface design issues, argues in a similar fashion that constructivist notions of learning as activity, exploration, and creation are well suited to the computer environment. His view is that traditional education is passive, and that computers offer an opportunity for engagement that is powerful and new. Shneiderman claims that the constructivist approach to computer learning is very different from the behaviorist teaching machines, computer-assisted instruction, intelligent computer-assisted instruction, and intelligent tutoring systems. The constructivist view, based to a large extent on Piaget's theories, concentrates on interactive learning environments and discovery learning.

VYGOTSKY AND COOPERATIVE LEARNING THEORY

A common criticism of Piaget by educators is that he did not give enough attention to social aspects of learning and emphasized the notion of the solitary learner too much. There is evidence that Piaget recognized this oversight in his early work on children. In fact, later in his career he spoke of social influences as one of the factors in human development. Nevertheless, for Piaget, behavioral conformity is not the same as understanding. He warned that social forces can lead to rigidity of thought rather than creativity. Finally, for Piaget, social experience and education are necessary, but not sufficient in themselves (Piaget, 1995).

Lev Vygotsky, the Russian theorist, has become popular to educators looking for an approach that emphasizes social aspects of learning more than Piaget seems to do. Taking a slightly different approach concentrating more on the importance of social interaction for learning, Vygotsky (1997) argued that individual consciousness is built from outside through relations with others. He flips Piaget's model around and states that the true direction of the development of thinking in children is not from the individual to the social, but from the social to the individual.

Following further in the theoretical lineage of Vygotsky, much has been written about cooperative learning. Slavin (1983) notes that research on cooperation in education has been conducted since the beginning of the 20th century, and research on the use of cooperative methods began in the early 1970s. However, he notes that none of it is very extensive. According to Slavin, in practice, students in traditional classrooms are rarely allowed or encouraged to help one another. The incentive system puts students in competition with one another and thus discourages collaboration. He argues that the essential ingredient of cooperative behavior in educational settings is the attempt of each to facilitate attainment of the goals of fellow students. However, Slavin claims that the approach of students helping each other by itself doesn't seem to be an advantage to students. Additionally, assigning grades for group work is a problem because it sometimes leads to trouble when weaker students are not motivated to participate.

Consequently, research shows cooperation can have both positive and negative effects, depending on how it is structured in the classroom. One effective strategy is to have both individual and group scores for collaborative projects. Thus, part of a grade is for individual work, and part for group work. Slavin argues that incentive structures, not task structures, are what make cooperative learning effective. The incentive to help each other causes the increase in learning, not the efficiency in task structures gained by teams. Cooperative learning methods high in individual accountability of groups members are more likely to produce greater learning. Positive effects of cooperative learning are seen in self-esteem, peer support for achievement, and an internal locus of control. Additionally, Slavin claims that research shows that special needs students can benefit from cooperative learning environments by overcoming friendship and interaction barriers. We will see later in Chapter Three that cooperative learning methods have been shown to be particularly effective in distance learning environments. The background for its use comes from the emergence of Vygotsky's theories in the last few years and their application in the classroom.

SKINNER

B.F. Skinner was one of the first behavioral psychologists to consider the use of computers for educational uses. In *The Technology of Teaching* (1964), Skinner argues that the use of technology in teaching can increase learned behavior by organizing learning objectives, increasing the frequency of positive reinforcement, customizing the learning experience, and freeing teachers from repetitive teaching. He focuses on teaching as the structuring of opportunities for reinforcement: "...teaching is simply the arrangement of contingencies of reinforcement" (Skinner, 1964, p. 5). Skinner further defines this learning opportunity as the environment where learning takes place, the occasion when behavior occurs, the behavior itself, and the consequences of behavior. His famous teaching machine is essentially a reinforcer: "The application of operant conditioning to education is simple and direct. Teaching is the arrangement of contingencies of reinforcement under which students learn" (Skinner, 1964, p. 64). Skinner also points out the importance of the effective scheduling of reinforcement in educational design in terms of frequency and type.

Overall, Skinner sees the key advantage of teaching machines in that the user has immediate feedback from the machine: "The important features of the device are these: reinforcement for the right answer is immediate" (Skinner, 1964, p. 24). Constant interchange between program and student much like a tutor is the goal. "Unlike lectures, textbooks, and the usual audio-visual aids, the machine induces sustained activity" (Skinner, 1964, p. 39). Skinner promotes the notion of teaching machine as tutor in pacing students through appropriate level material and through prompting, hinting and suggesting ways students can arrive at correct answers. In addition, Skinner points to the advantages of teaching machines beyond issues of behavior modification that allow improvements in class management, asynchronous learning, and customization.

LEARNING STYLES, MULTIPLE INTELLIGENCES, AND SELF-REGULATED LEARNING

Recently, much has been written about learning styles that may have particular relevance to the customization of educational software for learners. The notion of "learning styles" came about when differences were found in the organization and processing of information by students. Learning style generally refers to learning dispositions that students adopt in educational environments and is also sometimes called "learning approach" and "learning orientation." Howard Gardner (1993) proposes the related notion of "multiple intelligences" that has gained wide-spread acceptance among educators. He sees seven basic intelligences: musical, bodily-kinesthetic, logical-mathematical, spatial, linguistic, interpersonal, and intra-personal. Since his original definition of multiple intelligences, Gardner's research has evolved to define more specific intelligences.

In recent years there has been increasing debate about the benefit of this notion of learning styles and multiple intelligences in higher education. Generally, these theories present a move away from the classification of learners, towards a classification of learning instead. Some research shows that when introduced to learning environments suited for specific kinds of learning well suited to them, students perform better (DiPaolo, 1999). This ability to customize learning to suit an individual learner's style may be a key advantage of computer-based learning.

In a regular classroom, teachers are forced to use generic approaches to learning style, while in distance learning teachers can often customize the learning experience using those methods best suited for students.

"Self-regulated learning" is related to learning styles and multiple intelligences theories. According to Pintrich (1995), self-regulated learning attempts to control behavior by having individual students, not teachers, set goals. It involves active, goal-directed self-control of behavior and cognition. Pintrich claims that all students can learn to be self-regulating, but that it is particularly appropriate at the college level. A significant aspect of self-regulated learning is awareness of learning style through self-reflection. In contrast to informal learning situations, self-regulated learning requires formal self-monitoring to outcomes and reflection on outcomes. Pintrich argues that grades work against self-regulated learning, and that an orientation towards mastery of course material is more productive.

The question that arises in relationship to the research literature on learning styles is: How can learning styles be accommodated in computer-based learning environments? Gilbert and Han (1999) found that a computer program could be created that used different approaches to instruction based on student scores on computer learning tests. If students performed well, the software continued with a certain style. If they performed poorly, another approach was attempted. This general attempt to conform to student attributes may become more sophisticated as we better understand these tendencies.

ADULT LEARNING THEORY

DeMartino (1999) argues that while currently the literature is dominated with the debate between constructivism and behaviorism, adult learning theory might be the most important to look at given the adult population using distance learning. There is a lot of validity to this observation. For many reasons, including the need for convenience, the higher level of foundational knowledge, and other characteristics of adult learners, distance learning is likely to be more useful to adults than children. What does the research literature on adult learning tell us that may be useful for educational software designers?

Carl Rogers and Abraham Maslow are probably two of the most important influences on the field of adult learning. In *Freedom to Learn* (1968), Rogers, the humanistic psychologist, describes his theory of experiential learning characterized by personal involvement, self-initiation, pervasiveness, and evaluation by the learner. For Rogers, the essence of learning is personal meaning built into the classroom experience. He attacks the notion of teaching and instead emphasizes the importance of the teacher as a facilitator. Rogers suggests that the goal of education should be the facilitation of change and learning. He stresses process rather than static knowledge and the importance of learning how to learn. Rogers suggests that self-discovered types of learning are the most meaningful and the longest lasting. He points out that it is important for learners to feel valued in order to learn. Rogers proposes that the goal of education must be to develop a society whereby people become more comfortable with change than with rigidity. He finally points out the effectiveness of the group as a vehicle for facilitating constructive learning, growth and change.

Maslow (1970) presents a notion of motivation combining James, Dewey and Gestalt psychology that he calls "holistic-dynamic theory." He begins by looking at needs and "self-actualization" and proposes that curiosity and the need to learn and have competency in the world are instinctual. Maslow argues that learning is change in personal development and character structure or movement towards self-actualization. Maslow criticizes traditional education because it fails to give individuals the opportunity to examine reality directly with a fresh, individual perspective.

More recently, Merriam and Clark (1991) look at how the major parts of adult life (work, love, and learning) are woven together in healthy individuals. Using case studies of more than 400 subjects, the authors identify three patterns of work and love in adulthood: the parallel pattern, the steady/fluctuating pattern, and the divergent pattern. They show how common experiences such as parenthood and divorce shape adult identity and the role that learning plays in each pattern formation. In their study, the authors find that learning plays a key role in development of the capacity to work and to love. Their results show that positive changes in work tend to coincide with upturns in love relations and an increase in informal and formal learning.

In the last thirty years a great deal has been written on the subject of adult learning in a general way. Malcolm Knowles, Patricia Cross, and Stephen Brookfield are three of the major theorists in the field. Knowles (1979) uses the term "andragogy" as opposed to "pedagogy" to argue that adults learn in ways different from children. He claims that adults have different motivations, goals, and expectations in regard to education. Adults are more self-directed, can be a rich resource for learning because of their life experience, can immediately apply what they learn, and are more problem centered. Consequently, the teaching methods for adults need to be different. His andragogy emphasizes guiding and facilitating over teaching. Borrowing from the work of Rogers, Maslow and Dewey, Knowles promotes learner-centered education and the importance of personal inquiry. The learning climate should be informal and collaborative, involve mutual planning, self-diagnosis, learning objective planning, and assessment. The instructional materials should be scheduled by learner readiness, focused on questions, and use an inquiry method. Further, Knowles argues that it is no longer realistic to define education for adults as the transmission of what is known. Instead, the main purpose of education should be to develop skills of inquiry in the learner. Thus, one can see the importance of self-directed learning as a technique.

Cross (1981) looks at the growth of the learning society and its characteristics in terms of motivation and learning theory. She describes the increase of mandated continuing education for adults and other pressures driving the phenomenon of lifelong learning. In analyzing motivation, Cross cites statistics showing that adults continue their education primarily for pragmatic reasons. She also looks at barriers to learning and finds that costs, lack of time, and past unpleasant experiences in education are primary factors. Cross analyzes research on specific populations in continuing education and argues that upwardly mobile lower-middle classes with an emphasis on mobility and status and a concentration on satisfying belonging needs within the nuclear family are much more interested in continuing education than what she calls lower-lower classes. She sees two models for continuing education: self-directed learning, and traditional classroom learning with varying degrees of low-threat to high-threat learning situations. Much as we saw with Merriam and Clark, Cross points out the effect of life transitions on adult motivation to learn. Learning in relationship to adult development centers on issues of reduced physical abilities and debate over loss of intelligence. Cross focuses her argument on the question of identifying the age of diminished capacity and the importance of learning speed in education. She notes that research shows individuals are remarkably consistent in learning method tendencies, and that some retain intelligence very consistently throughout adulthood. Cross criticizes Knowles for his theory of andragogy as being difficult to implement. In looking at the humanism versus behaviorism debate, she sees value in both for adult learners. She puts forward her own model called Characteristics of Adults as Learners (CAL), pointing to the primary difference between adult and child learners as being the part-time and voluntary nature of adult learning.

Brookfield (1986) analyzes the current literature and practices of adult education and andragogy, which he argues is more a group of assumptions about adult learners than a proven theory of adult learning. He points out that adult learners as a group are a very complex set and that they are consequently difficult to generalize about. In looking at self-directed learning theories, Brookfield argues that experience has shown that many adult learners do not respond to self-directed learning courses. In fact, in contradiction to the positions of both Knowles and Cross, most adult learners do not want self-directed learning and function poorly in such courses. Furthermore, Brookfield points out that some of the theorists of adult education have gone too far in allowing students to dictate the direction of the learning process. He also holds that much of andragogy includes learning theories that are appropriate for children, not just adults. Furthermore, Brookfield denies the common perception that adult learning should be skill oriented and practical. His research shows that adults are often interested in studies that are impractical and more focused on personal development.

Others have concentrated on the more narrow field of professional education. Argyris and Schon (1974) look at professional education and its ineffectiveness. They describe the way professional education is often shunned by universities, and thus a theory of practice hasn't been properly addressed. Argyris and Schon propose that the purpose of professional education is not only to teach technique, but to teach methods. In another study, Schon (1987) looks at educational practices in professional schools and argues that design studios offer a model for education based on the reflective practicum. Reflective practicum is a process whereby students reflect in action while doing and also involves an on-going dialogue with a coach. Schon examines the practices of an architectural design studio and the professional clinical training of a psychologist, and reflects in particular on the relationship between the coach and the student as they work their way through real-life learning experiences. He argues that real life presents problems that are much more complex and ill-defined than problems in traditional educational environments. Schon argues that the practice of professionals involves a type of "artistry" including problem framing, implementation, and improvisation— all of which cannot be taught in traditional fashions. He proposes a model based on the freedom to learn by doing, with access to coaches who initiate students in the tradition of a professional calling. His notion of "reflection in action" involves on-the-spot experimentation. Schon focuses on the paradox of learning in professional education where students can only begin to understand something they are unfamiliar with by doing an activity in the practicum experience.

In general reviews of the field (Peters & Jarvis, 1991), questions about how psychology has influenced the articulation of principles for adult education are asked. What areas of psychology can best guide and inform adult education practice? Wolf (1993) connects adult development to issues of adult learning. She looks at the notion of "wellness" and argues that older adults would benefit from self-examination and education about the developmental process. Wolf claims that education can play an important role for older adults by offering an opportunity to both reflect upon their own lives and to understand the developmental processes.

As many of the adult learners integrate their educational experience with their work, workplace learning has also become a subject of research. Baskett (1993) examines self-directed learning in the workplace. The author describes the need for rethinking workplace learning in the face of increased demands for upgrading the skills and knowledge of workers, particularly so-called "knowledge workers." In reviewing the research literature, Baskett notes that the majority of workplace learning occurs informally and that it is difficult to isolate such learning experiences because they are embedded in the work. Examples of such learning include interactions in teams and committees, with both internal and external individuals, that involve problem solving and overcoming various human problems. Baskett and Morris (1992) report on a research project on how to enhance self-directed workplace learning and argue that most workplace learning occurs informally through interaction with both internal and external work relationships. They then identify factors leading to the encouragement of workplace learning including: continuous improvement as an organizational strategy, encouragement of employee involvement in decision-making, rewards for learning, organizational values paralleling those of the individual, managers who set learning examples, valuing differences, supporting risk taking, effective communication systems, the encouragement of collaboration and the formation of teams, and the creation of systems to support creativity and innovation.

Smith and Marsiske (1994) present a framework for looking at performance in the workplace from a life span developmental perspective. Research reviewed by the authors shows that aging lacks uniformity and has both positive and negative influences on performance specific to each individual. Similar to Cross, they find that adult intelligence through the lens of a life span perspective reveals that skill levels are very stable. Both mechanical and cognitive skill levels are remarkably stable over a life span. In fact, research shows that age has no apparent connection to performance. When there is decline in skills, rapid response and overall speed are generally the areas impacted. Nevertheless, some individuals show areas of selective growth as they grow older, including into old age. Workers who upgrade their knowledge bases tend to have more on-the-job experience to make such education relevant and experience against which to test new ideas. Smith and Marsiske suggest that increased knowledge of specific work-related domains represents the greatest educational opportunity. The study concludes by arguing that notions of workplace skills need to be broadened to include interpersonal communication skills, motivation, and experience in specific job-related domains. Furthermore, the authors suggest that more attention should be paid to informal learning opportunities in the workplace and to contexts that facilitate adult learning.

Covey (1995) emphasizes that lifelong learning is most effective in short daily study sessions and small doses of relevant on-the-job training. He argues that learning and development should be motivated by a desire to be of greater service. Covey estimates that twenty percent of the workforce is obsolete because of outdated skills. He claims that corporations cannot compete without creating a culture of continuous learning, with knowledge workers who are continually enhancing skills and updating technology. Covey believes that individuals, not companies, must take personal responsibility for professional development.

Confessore (1996) argues that learning must be integrated into all aspects of the job because of the need for increasingly skilled workers that traditional education cannot supply in rapidly enough, and because of the inability of companies to afford down-time while the worker is training. What has evolved is a need for workplace environments that integrate learning with every aspect of daily activity. Confessore sees reflective practice, action learning, self-directed learning, and learning organizations as becoming increasingly important in the workplace. Reflective practice requires the individual to consider the effectiveness and consequences of actions taken during work. He notes that all of these methods incorporate notions of increased learner control and an unbundling of time and place requirements. Confessore links self-directed learning to learning organizations. Behaviors identified as key to learning organizations are very similar to those of self-directed learning, including openness to alternatives, ability to make connections between seemingly disparate issues, topics and information, creativity, including flexibility and willingness to take risks, and personal efficacy. The capacity of a learning organization to adapt quickly and effectively is directly dependent on how well employees learn. Confessore argues that pervasive, natural and informal learning atmospheres are the most effective.

Writing about workplace learning, Brookfield (1987) looks at the nature of critical thinking and argues that it is not an academic conception, but that thinking critically is one of the important ways one becomes an adult. He analyzes the workplace as a resource for thinking and learning beginning by pointing out that research in this area has typically focused on the manager as someone who can learn on the job, while almost completely disregarding the lower-level workers. Classroom definitions of critical thinking have led to a lack of understanding of its importance in the workplace. However, Brookfield points out that many of the managerial activities directly involve critical thinking such as strategic planning, problem solving, decision making, leadership, entrepreneurial risk taking and team building. In terms of application to the general workforce, Brookfield describes attempts to create workplace democracy that gives frontline workers an opportunity to develop critical thinking skills. However, he argues that workplace democracy needs to include development of individual workers in terms of self-confidence and competency in order to succeed. Finally, Brookfield points out the importance of the use of critical thinking skills in the formation of a positive self-image.

The research literature includes references to various training models for workplace learning. Cusimano (1995) describes a training approach developed by Manufacturing Technology Strategies to help companies develop and empower employees to become knowledge workers at the blue-collar level. The adaptable function-oriented training system teaches blue-collar workers how to solve problems, write technical manuals, and become responsible for their own learning. The approach involves understanding the technology and equipment, conducting a needs analysis, and developing a training curriculum customized to each worker's needs.

Malloch, Cairns and Hase (1998) review two complementary research projects taking place in Australia and the United Kingdom examining the implications of the Capability Learning Model in the workplace. The Capability Learning Model is defined as the application of both abilities and values within varied and changing situations to better formulate problems and actively work towards solutions as a self-managed learning process. The authors emphasize that this approach provides a situated learning experience, that learning can be effectively situated in the workplace. Additionally, facilitation rather than training should be the approach, and an emphasis on learning through the work community is important.

Montgomery (1996) describes a Career Development Model (CDM) as a new organizing structure for Andersen Consulting. The CDM implements research on integrative learning by identifying skill tracks, specialized skill domains, and levels of competence. They used a process of analyzing their work and learning needs including doing, looking, thinking, continually evaluating relevance, and planning. Montgomery and Lau (1996) describe a model developed by Andersen Consulting for a new approach to workplace learning called an Integrative Learning Model. The authors describe the changing nature of business where learning has become a major function. As others have commented on above, they found that learning in the workplace is most effective when it is informal and when it takes place in the real workplace environment. Conversely, research shows that artificial settings and formal training is a barrier to workplace learning. The authors describe the six-part make-up of their model as: 1) to access new information, ideas, experiences, or perspectives; 2) to integrate life experiences; 3) to utilize a reflective learning approach; 4) to apply new concepts within temporarily transformed experiences; 5) to apply new knowledge; and 6) to evaluate relevance and the value of new experiences. The authors examine the notion of integrative learning, which they characterize as used to build knowledge and skills in a safe environment using coaching techniques, when appropriate, teams, and trust in the process.

One can see from this review of the adult learning research literature that there are many possible ramifications on software design. The part-time and solitary nature of much of adult learning and the overall motivation structure are fundamental elements to understanding effective design. Additionally, the research on workplace learning might be one of the most important areas to consider. How best can software be designed to facilitate workplace learning? How can informal learning be encouraged and structured using computers?

CONCLUSION

One could take either side of the behaviorism versus constructivism debate in relationship to technology and argue successfully. Clearly, each approach has advantages for certain kinds of skills to be learned and certain types of students. However, the technology cannot be separated from the theory. One problem with technology-enabled learning is that it is sometimes used to automate unsuccessful teaching strategies. Ineffective classroom lectures posted to an online course will not improve learning. The technology of education will only be successful to the degree it uses effective learning theories. Will technology lead this change, or will improved teaching strategies come first? The indications here are that they will occur together, with technology offering the occasion to rethink teaching and learning.

Although instructional design has been traditionally dominated by behaviorist approaches to skill development, in recent years constructivism has begun to make inroads. What are the implications of constructivism for instructional design? In many ways, technology fits well with constructivist, learning-centered theories. The techniques of concept mapping, computer programming, rich simulated worlds, and hypertext all make sense in the constructivist paradigm. The constructivist framework puts an emphasis on assisting the student in constructing his or her own knowledge through the use of computers. An active learner elaborates upon and interprets the information presented in an instructional program. A constructivist approach need not only be of the discovery learning type, but can also focus on more direct instruction, as long as the emphasis is on going beyond the information given. For constructivists, context is an integral part of meaning. Consequently, constructivists propose working with concepts in rich computer environments that lead to seeing complex interrelationships. Constructivists believe that when learning occurs in isolation as separate topics, the learning remains inert. The goal is to create a computer environment where tasks take on meaning in a larger context.

Critics of Skinner's position on technology point to his overemphasis on changed behavior as learning. In my view, his methods are useful in utilizing repetitive learning methods, but don't begin to get at the real potential of a teaching machine. Skinner missed the real value of educational technology because he focused on reinforcement of behavior rather than the potential of a new learning medium that can assist deeper and more creative thinking.

Critics of the application of constructivism to educational technology raise problems such as the effectiveness of constructivist instruction when it tries to cover too much material, the lack of concern with the skill level of students, and the reliability of evaluation methods. Furthermore, contrary to the constructivists, some argue that there are different organizations of knowledge required to promote different learning outcomes, and that the learner need not always be in control. Are the demands that a constructivist learning environment place on the learner too great? Part of this issue has to do with what learning approach students are accustomed to using. Obviously, students accustomed to rote learning who are suddenly introduced to a more open constructivist learning environment are going to have to make adjustments in their thinking.

As Brookfield (1986) points out, much of the literature contains assertions about adult learning with little research to back them up. Overall, the complexity of the field, with its huge variation in the characteristics of adult learners that make them as a group very difficult to generalize about, is not fully appreciated. Future research needs to focus on this complexity and avoid reductionism that simply tries to create distinctions between adult and child learning without real supporting evidence.

What is the relationship between work and learning? Learning and work are directly tied through adult development issues of motivation, self-image, and by the changing nature of organizations focusing on knowledge as a primary resource. How do conceptions of the knowledge worker and the learning organization fit with notions of adult development? Learning organizations must be concerned with the individual development of each worker because it directly impacts productivity. The question therefore remains: how specifically do organizations encourage adult development in a systematic manner?

This chapter reviewed the relevant learning theory that directly relates to educational software design. In the next chapter I turn to one of the specific methods of teaching/learning in computer environments–the tutorial method.

Chapter III

Tutorial Method: Profiling, Customization and Agents

I felt like I had my own tutor (anonymous distance learning student, 2001 survey).

The two major approaches to distance learning are group and tutorial. In this chapter I concentrate on individual tutorial practices and principles applied to computer educational environments. One of the most important areas of research in distance learning pedagogy focuses on the ability of computers to customize the learning experience to meet specific learner needs. This customization aspect of distance learning involves structuring the course content in personalized ways through the use of computer agents and other forms of artificial intelligence.

Respondents to the 2001 survey indicated that they feel that one of the roles of effective distance learning is to provide customization. In the following figure we see that 73.2% of the students surveyed responded strongly agreeing or agreeing to the statement, "It is important that courses are customized to meet my specific learning needs and to adjust to my learning style."

Another possible role for computers is to act as automated tutors or learning assistants. The 2001 survey showed that students had a strong interest in having the computer act as a learning assistant. Students responded 89.6% of the time either strong agreement or agreement with the statement, "I would like to have the computer serve as a learning assistant or agent in distance learning courses."

Figure 3: Customization Importance (Questions 1 & 9)

Delivery format * It is important that courses are customized to meet my specific learning needs and to adjust to my learning style. Crosstabulation

			It is important that courses are customized to meet my specific learning needs and to adjust to my learning style.				
			strongly agree	agree	disagree	strongly disagree	Total
Delivery format	computer-based	Count	12	31	15	1	59
		% within Delivery format	20.3%	52.5%	25.4%	1.7%	100.0%
	videotape	Count	16	29	16	1	62
		% within Delivery format	25.8%	46.8%	25.8%	1.6%	100.0%
	correspondence	Count	1	3	1		5
		% within Delivery format	20.0%	60.0%	20.0%		100.0%
	other	Count	1				1
		% within Delivery format	100.0%				100.0%
Total		Count	30	63	32	2	127
		% within Delivery format	23.6%	49.6%	25.2%	1.6%	100.0%

Figure 4: Computer as Learning Assistant (Questions 1 & 22)

Delivery format * I would like to have the computer serve as a learning assistant or agent in distance learning courses. Crosstabulation

			I would like to have the computer serve as a learning assistant or agent in distance learning courses.				
			strongly agree	agree	disagree	strongly disagree	Total
Delivery format	computer-based	Count	19	23	4	1	47
		% within Delivery format	40.4%	48.9%	8.5%	2.1%	100.0%
	videotape	Count		1			1
		% within Delivery format		100.0%			100.0%
Total		Count	19	24	4	1	48
		% within Delivery format	39.6%	50.0%	8.3%	2.1%	100.0%

While many argue for the pedagogical advantages of group interaction, for some students this is not preferred. In the following responses from the 2001 survey, students expressed this preference for individual study.

Do not need interaction with others, want to do independently (anonymous distance learning student).

I did not spend a lot of time/energy reading other students' work samples—they were not valuable to me (anonymous distance learning student).

One student expressed a desire for only an advisor relationship with the faculty member.

I wish more classes of this nature were offered with an "advisor" for questions (anonymous distance learning student).

Another student concentrated on the quality of the instructional materials, not the interaction with the faculty or other students.

> Materials make all the difference...Video very clear...I did completely on my own, must [have] called instructor once, and she handled it—never met her (anonymous distance learning student).

Rather than the group interaction of a traditional classroom, students often want a tutorial kind of relationship. In the 2001 survey, more than half (50.5 percent) indicated "agree" or "strongly agree" that they wanted a one-on-one tutorial experience.

Figure 5: Students Wanting Tutorial Structure (Questions 1 & 17)

Delivery format * I would like the distance learning format course to be structured like a tutorial with one-on-one contact with a tutor. Crosstabulation

			I would like the distance learning format course to be structured like a tutorial with one-on-one contact with a tutor.				
			strongly agree	agree	disagree	strongly disagree	Total
Delivery format	computer-based	Count	2	19	26	1	48
		% within Delivery format	4.2%	39.6%	54.2%	2.1%	100.0%
	videotape	Count	6	21	19	1	47
		% within Delivery format	12.8%	44.7%	40.4%	2.1%	100.0%
	correspondence	Count	1	1	3		5
		% within Delivery format	20.0%	20.0%	60.0%		100.0%
	other	Count		1			1
		% within Delivery format		100.0%			100.0%
Total		Count	9	42	48	2	101
		% within Delivery format	8.9%	41.6%	47.5%	2.0%	100.0%

Administrators responding to the 2000 survey also focused on the tutorial nature of the distance learning experience responding 73.2% of the time that they agree or strongly agree with the statement, "Courses are like one-on-one tutoring with the faculty members, providing rich and prompt feedback to the students."

Figure 6: Courses Like One-on-One Tutoring (Question 45)

Courses are like one-on-one tutoring with the faculty member, providing rich and prompt feedback to the students.

		Frequency	Percent	Valid Percent	Cumulative Percent
Valid	strongly agree	43	24.4	25.6	25.6
	agree	80	45.5	47.6	73.2
	disagree	37	21.0	22.0	95.2
	strongly disagree	8	4.5	4.8	100.0
	Total	168	95.5	100.0	
Missing	System	8	4.5		
Total		176	100.0		

Tutorial Method: Profiling, Customization and Agents 31

In the interviews for the 2000 survey, some administrators spoke about tutorial relationships created for their distance learning courses. Tutorial or mentorships differ from group learning in that they concentrate on the teacher-learner relationship, rather than the communication among students. Furthermore, this method alters the role of the instructor in a formal way. In the following excerpt we see how one administrator structured a mentor program based on the British Open University method.

> The Open University of the United Kingdom uses tutors–they call them tutors. On their main campus, there are no students, it is a production center. Then they have regional centers, thirteen I believe, maybe more. Those regional centers are a hub that go out to a series of study centers. Study centers are places where students can meet face-to-face with tutors, or not. They don't have to. They can communicate by computer, phones, whatever. The materials are designed so that one can work through them on his or her own. They have the safety net of the tutors. The tutors also do all of the grading and marking, and they send that into the central campus, and the lead faculty are responsible for the final grade. Now we took this model and use mentors instead of tutors and they are responsible to manage and coordinate groups of students within a section of a course all online. Our mentors are all over the state, some are even out of state now, and are used in our distance learning courses. Our two plus two online program is set up to recruit, train, and support mentors for our undergraduate degree completion programs. So those are the things we learned from the British Open University (Carole Hayes, Coordinator, External Relations and Development, Office for Distributed and Distance Learning, Florida State University).

The following administrator describes a similar system.

> Students are introduced to Saybrook's collaborative approach to learning at the Residential Orientation Conference. There they also are introduced to faculty, advisors, and administration, they meet and network with other incoming students, and learn how Saybrook's distance learning process works. Relationships are formed and continue to be developed at Residential Conferences (RCs) that are held twice a year. Ways in which students can learn include: one-on-one mentorship relationships with faculty; as a small cohort of students; courses that are online; workshops or seminars completed at the RCs; independent studies; peer learning; case-based learning; and practica and internships.

> Courses that are available have a learning guide (an extended syllabus), a course reader, and required and recommended readings. Usually, by conferring with faculty, a course can be tailored to meet the needs of the individual student. Most three-unit courses include the writing of three ten-page papers which are transmitted by email to faculty for review and comments. Faculty function as mentors or advisors in this model (Kathy Wiebe, Admissions Coordinator, Saybrook Graduate School).

One administrator spoke about the emphasis placed on tutor-student interaction.

> We have the software and hardware people. I refer to my people as the "warm wear." Our point is to make it real, make it happening, alive, dynamic for students. Mentors are very proactive in taking the initiative to engage students. The average class size is 15 to 20 students, so at the beginning of the term the mentor needs to contact each student and get a response. If there isn't a response, we give them lots of information about how to go about getting one. And then it is their responsibility on a weekly or bi-weekly basis to initiate contact with those students and make sure that they know the mentor is the first point of contact, to counsel the students. And I don't mean counseling in the academic sense. We have academic counselors. But our demographic is so varied that the mentors need to understand the characteristics of the adult learner, and understand how to help with the new technology, to set up a new schedule in their lives, all those sorts of things. So there's a living contact point (Carole Hayes, Coordinator, External Relations and Development, Office for Distributed and Distance Learning, Florida State University).

An administrator from a specialized institution reported on a model designed to fit individual needs.

> Saybrook began thirty years ago, to provide education that combines residential conferences, distance study, and individualized mentorship from renowned leaders in the humanistic field. Over the years we have incorporated the use of the computer and email for sending papers and holding discussions with individual faculty or in the case of some classes that are online, and for expediting registering for classes and other administrative procedures, but the basic model, which encourages mentorship and individualized learning, still holds (Kathy Wiebe, Admissions Coordinator, Saybrook Graduate School).

One administrator spoke of the challenge to screen students through automated self-tests to make sure that they have the temperament and skills to learn in a tutorial model.

> WebCT is what we use and there is a tutorial that goes along with that. Our office answers a lot of questions about that. We have on our Web site a self-test that we ask our students to look into. And of course some of the students that need to, don't. Are you a self-directed person? Those kinds of things. We say right there that if you answer, "No" to a certain number of questions, this might not be the best idea. Then again because we are an open door college, we cannot prevent people from registering for a class (Thornton Perry, Director of Distance Education, Bellevue Community College).

A major issue in tutorial models is how to help prioritize and stress areas within a broad subject matter. A student responding to the 2001 survey indicated that one of the biggest problems with videotape courses with little direction from a faculty member was in making judgments about importance and to focus attention.

> My biggest problem with the videotape courses was the lack of direction on which topics are most important. In a classroom environment a professor guides the student on what is important and what is valuable (anonymous student in distance learning course).

Furthermore, one administrator pointed to the role of faculty in providing clarification for course material because interaction is more difficult in computer-based environments.

> When students start the distance ed[ucation] program here there are two things that I tell students. I took a couple of the courses myself when I started here, one of the distance ed[ucation] programs. Distance ed[ucation] courses are more difficult for two reasons: one, there is less interaction with the live faculty member. "I'm sorry, I didn't get that could you repeat it"—no problem, rewind the tape, that's easy. But amplification, clarification, implementation questions are more difficult in a distance ed[ucation] course (Jon Raibley, Assistant Director of the Center for Lifelong Learning, Western Seminary).

Fox (1993) points out that little research has been done on human-to-human tutorial communication. If one of the major ways distance learning is structured is through tutorials, clearly this is an area in great need of further research.

Fox's main theory is that the function of tutoring is the collaborative contextualizing of abstract symbols and problems, the situating into local problems for the student. This is verified by the comments of students in the 2001 survey (as seen above) who complain about the need for a tutor to help prioritize the content. Fox points out that the openings of tutorial sessions are particularly interesting because they reveal how the contextualizing activity is negotiated between teacher and student. Furthermore, it was found by Fox that both the student and teacher come to tutorial sessions without preconceived plans and that generally it is the student who initiates the process of planning. A plan is generally developed between the two parties, as well as an understanding of how much tutors will intervene in the discussion and problem solving. The joint contextualization of problem statements is a form of collaborative cognition. Help with the deep conceptualization of the problem is one important role the tutor plays. Fox found that in an analysis of the tutor-student discussions that students use a running commentary to describe their thought processes for the tutors. The tutors then intervene at various points to either correct or comment on the commentary. She found that tutors monitor not only what students say, but how they say it in order to better evaluate and assist.

SPECIFIC TECHNIQUES: USE OF EMAIL, AGENTS, TEXT FILTERING, PREDICTIVE TEXT GENERATION, CREATIVITY TOOLS

A number of computer-based techniques have been developed to assist with a tutorial orientation to learning. They include email, agents, predictive text generation, and creativity tools. We saw in the previous chapter that there is extensive and growing research literature on learning styles and multiple forms of intelligence. This literature points to the need to pay attention to individual differences in learners in order to increase learning. It may very well be that one of the most practical advantages of the use of computers in educational environments is in the ability to customize learning experiences for individuals. Nevertheless, the research literature on the customization of learning in computer environments is limited. Sumner and Taylor (2000) claim that a personal learning manager supports dynamic updates, reuses other courses and materials and is annotatable. They point out that many programs need to provide ways for users to interpret the new media, much like explicators were used at the turn of the century to stand in front of the cinema screen interpreting the action. Furthermore, Sumner and Taylor argue that providing ways to customize software is important for special needs populations. Trying to accommodate special needs students can benefit all students through customization of media.

Students often report that contact with instructors is freer via email than by telephone. Although tutors tend to spend more time on email interactions with students than on the phone, they often still prefer it because of the control gained. Furthermore, tutors find that they can re-use sections of correspondence to reply to common inquiries and responses. Also, email communication is much faster than mail. Petre, Carswell, Price and Thomas (2000) found that email responses by instructors to students work was reduced from 5-7 days through mail, to 2-3 days by email. They claim that tutorial interaction through email is best when humor is used. Witty problems, lightly-phrased coaching and criticism, clearly, tasks structured, milestones and a review of key points in the material are productive techniques.

Probably one of the most exciting tools for learning for the future is in the development of intelligent tutors. As many have noted, the application of Dewey's educational philosophy puts an enormous load on the teacher, one that is impractical for broad-based implementation. Computers have the potential of meeting this need for labor through the development of intelligent tutors. Many of the programs thus far developed as tutors in training applications have been behaviorist in orientation. Partly, this is because of the limitations of the software itself. However, as research in computer agents and artificial intelligence advance, tutors will become more sophisticated.

One way this is already changing is in the area of intelligent agents for research. Agents are active and ever-present software components that perceive, appear to reason, act, and communicate (Huhns & Singh, 1998). According to Weiss and Dillenbourg (1999), agents have four components: sensor, motor, information, and reasoning. Agents, sometimes also referred to as guides and personal assistants, first appeared in the form of travel agents helping users to make their way through applications (Oren, Salomon, Kreitman & Abbe, 1995). The key aspects of agents are anthropomorphic presentation, adaptive behavior, multi-modal, dialogue based, ability to work with vague goal specification (mixed initiative), supply what you need, and the ability to work unattended (Shneiderman, 1995; Huhns & Singh, 1998). Also, agents suggest a natural way to present multiple voices and points of view (Oren, Salomon, Kreitman & Abbe, 1995) and can involve a degree of improvisation (Chapman, 1991). Many believe that human-human interaction is a good model for human-computer interaction (HCI) and consequently look to agents as a perfect HCI solution (Shneiderman, 1997).

Agents are viewed in two extreme views that reflect on the degree of artificial intelligence used in their construction. One view sees agents as conscious, cognitive entities. The second major view is that agents are only programs responding to commands or command sets made in advance. Applications involving information access, filtering, electronic commerce, education, and entertainment are becoming ever more prevalent and have in common a need for mechanisms for finding, integrating, using, managing, and updating information, all of which agents are intended to perform. In recent years, much has been written on agents, as the trend has shifted from passive interfaces to active interfaces (Huhns & Singh, 1998).

Some designers promote the notion of adaptive and/or anthropomorphic agents who carry out the users' intentions and anticipate needs. The famous bow-tied, helpful young man in Apple Computer's 1987 video on the Knowledge Navigator, and Microsoft's unsuccessful BOB program, are examples of early attempts at anthropomorphic computer agents. Although the majority of the literature is highly optimistic about the promise of agents, some doubt that they will work because of the difficulties in understanding the context of information and the need for users to trust computer agents (Head, 1999; Shneiderman, 1995). Shneiderman (1995) argues that agents offer promise, but a good alternative to agents may be to expand the control-panel metaphor and establish personal preferences. Applied to the Internet, intelligent agents can track the tendencies of the user and then collect information that fits the user's interests.

Alison Head (1999) suggests that agents will be a big disappointment because they have trouble dealing with anything but the most linear type of information. What is missing in the intelligent tutor applications of present are the more advanced functions of a human teacher who can lead a student in new conceptional directions and make connections not readily apparent. While it may be true that artificial intelligence will never reach the point where it can serve as a tutor on the level of a human teacher, intelligent agents customized to the fit the user's interests may serve as useful tools approximating the function of a tutor.

Text filtering is another type of artificial intelligence often mentioned in HCI literature. Text filtering may be one of the functions of an intelligent agent and is an information seeking process whereby documents are selected for specific information needs (Shneiderman, 1997). Luhn is credited with identifying a modern information filtering system and introducing the idea of a "Business Intelligence System" in 1958, where library workers created profiles for individual users and produce lists of new documents for each user. Selective Dissemination of Information (SDI) became a field and resulted in the creation of the Special Interest Group on SDI (SIG-SDI) of the American Society for Information Science. By 1969, sixty operational systems were being used generally following Luhn's model. Denning coined the term "information filtering" and broadened a discussion that had traditionally focused on generation of information to include reception of information as well. He described a need to filter information arriving by email in order to separate urgent messages from routine ones and customize to the needs of the user. Malone introduced an alternative approach called social or collaborative filtering where a document is based on annotations to that document made by previous readers.

Predictive text generation is another form of artificial intelligence that uses a context-sensitive technique for enhancing expressive communication to suggest what the user might want to type next on the basis of preceding input. Predictive text generation is now familiar to many who use the latest Microsoft Office products. Many of the traditional uses for this form of HCI are for those with special needs. It works by accelerating typewritten communication with a computer system by predicting what the user is going to type next. Good touch-typists are likely to find predictive text generation a hindrance, but moderate to poor typists find it helpful, especially for highly structured text (Darragh & Witten, 1992).

A final use of artificial intelligence in HCI repeatedly described in the literature is for visualization and creative endeavors. In the 2001 survey it was found that students had a clear interest in using the computer for visualization and creative endeavors. Respondents indicated 89.1% of the time they either strongly agree or agree with the statement "I would like to use computers for creative and visualization purposes in distance learning programs."

Shneiderman (1999b) describes the need to support creativity for users as a challenge for HCI designers. His model called "genex" includes four stages focusing on collecting previous works stored in digital libraries, relating with peers and mentors at multiple stages, creating through exploration and discovery, and donating by disseminating the creative results to digital library collections. To this scheme Shneiderman adds visualization, free association, and replaying histories as areas of needed research. He sees visualization as supporting creative work by enabling users to find relevant information and identify patterns. Further, important aspects of computer assistance with creativity include constructing meaningful overviews, zooming in on desired items, filtering out undesired items, and showing relationships among items. Another approach is to use multiple coordinated views for exploring information creatively. Each view is a visualization of some part of the information, and views are tightly linked so that they operate together as a unified interface. Spotfire, Xerox PARC's perspective wall, Yale computer science

Figure 7: Computers Used for Creative and Visualization Purposes (Questions 1 & 23)

Delivery format * I would like to use computers for creative and visualization purposes in distance learning courses. Crosstabulation

			I would like to use computers for creative and visualization purposes in distance learning courses.				Total
			strongly agree	agree	disagree	strongly disagree	
Delivery format	computer-based	Count	14	26	4	1	45
		% within Delivery format	31.1%	57.8%	8.9%	2.2%	100.0%
	videotape	Count		1			1
		% within Delivery format		100.0%			100.0%
Total		Count	14	27	4	1	46
		% within Delivery format	30.4%	58.7%	8.7%	2.2%	100.0%

professor David Gelernter's LifeStreams, and LifeLines are other systems that take a similar approach to information exploration (Shneiderman, 1997).

Clifford Pickover (1991) is another of the main proponents for the use of computers as aids to the human imagination. He argues that computers are providing mankind with an unlimited unparalleled aid for the imagination. Pickover proposes visualization for scientific use through both simple and advanced computer graphics as a way to help understand complicated data. According to Jonassen (2000), visualization tools help learners understand concepts and principles for the interpretation of information rather than construction of knowledge. He sees intelligent agents as software acting on behalf of people, as personal research assistants.

CONCLUSION

In this chapter I looked at the first of two main approaches to distance learning—the tutorial method. We saw in the data from the two surveys that administrators use this method to a great extent and that students often view their experience in distance learning courses as tutorials. The research literature on attempts at customization of learning in computer environments to address learning styles was reviewed and found to be still in a very early stage of development. It is important to note here that many leaders in the industry (such as the University of Phoenix) are now examining how they can utilize research on learning styles to increase learning in computer-based classrooms. Clearly, this is an important area for future research. We then turned to an examination of how specific technologies are used in the tutorial method, such as email, text filtering, predictive text generation, and visualization tools. Questions that remain to be addressed in the second part of this book are how the tutorial method can best use the computer medium, and what specific issues arise using this approach with computers.

In the next chapter, group learning, the second main approach to distance learning, is examined. The distinction between these two methods is crucial to understand in both the administration and teaching of distance learning courses. Particularly in the United States where the adherence to group teaching methods appears to be great, it is important that administrators and faculty understand the differences in the two methods.

Chapter IV

Group Method: Virtual Teams and Communities

At this time, group or team learning is probably the clearest pedagogical approach to distance learning in the research literature. With a strong basis in learning theory, time and again those practicing and writing about distance learning have focused on learning through teams and discussed the dynamics of communication among students and faculty that occurs when this approach is taken. In this chapter, I examine the nature of community in traditional higher education, the research literature on virtual teams, and lessons learned in business in relationship to virtual teams applicable to distance learning.

The 2000 survey showed that most distance learning administrators (32.0% "strongly agree," and 52.1% "agree") distance learning administrators state that their courses offer an opportunity to collaborate with other students on projects.

Figure 8: Collaboration Among Students on Projects (Question 42)

Courses offer opportunity to collaborate with other students on projects.

		Frequency	Percent	Valid Percent	Cumulative Percent
Valid	strongly agree	54	30.7	32.0	32.0
	agree	88	50.0	52.1	84.0
	disagree	22	12.5	13.0	97.0
	strongly disagree	5	2.8	3.0	100.0
	Total	169	96.0	100.0	
Missing	System	7	4.0		
Total		176	100.0		

Copyright © 2003, Idea Group Inc.

In the interviews with administrators from the 2000 study, student collaboration was identified by many administrators as important. In the following interview, the administrator points out the difference between online and traditional correspondence courses as centering on the ability to collaborate.

> That is imperative. Most of the faculty require in online courses threaded discussions and/or chat rooms, and require that kind of participation. That is built into the software package we use. Correspondence courses [have] probably very little student-to-student interaction. So if that's a big part of a faculty member's pedagogy, then we can't develop those [correspondence] courses. And I'd say most of our correspondence courses are conducive to the isolated student working by themselves (John Burgeson, Dean, Center for Continuing Studies, St. Cloud State University).

Indeed, some respondents claim that a stronger sense of community is developed in distance learning courses than in traditional courses.

> Yes, if you talk to the students they tell you that this is more so than in traditional classrooms (Vice-President, anonymous large, independent, Eastern U.S. doctoral degree-granting institution).

Respondents spoke about techniques they use to form communities through various types of personal communication. One common way of personalizing the online experience is to have students post introductions and photos of themselves online.

> Yeah, it's one of the things I've worked hard at. Some of the things are small, but if you don't do it, you're not as good as if you do do it. That each student has a picture of themselves online, so that people can know what they physically look like. That each student has a short bio[graphy] of themselves online. The way that we have traditionally used that first hour of class, you know, "tell me one thing about yourself that no one else knows." Here you get beyond that, how many kids you have, what you do for a living, things that get beyond your bio[graphy]. But here you have a picture of yourself and a bio[graphy], and you can read it at will, not just the first hour of the class, before you get to know them. They say, "Gary's an interesting guy. I'm going to look at his bio[graphy] again." People know a lot more about each other than they did before we had them do the online bios. Blackboard has an area for this that they call their own

personal web page. One benefit for me is that I learn student names much sooner when I can connect a face to a name (Don Cardinal, Chapman University, School of Education Faculty member).

The use of chat rooms for communication amongst students was also identified by respondents as something that is commonly used, often as an option for interested students.

It is not required to my knowledge, but it is encouraged. Some instructors set up chat rooms, and other students set up groups on their own in whatever format they choose to do it in. It is encouraged, but not required to my knowledge by any of the programs (Allan Guenther, Marketing Coordinator, Distance Education, The University of Alabama).

One key teaching technique used in the courses is the use of group projects and tasks. The following excerpt describes the use of projects in different disciplines.

The Blackboard software we use provides for a great deal of flexibility. Say you have 100 people in a course with one lead instructor who designed the course and who is final authority on assessment. Then there are four to five mentors who are assigned specific students for their group. But within the software, say for instance a computer science course, the mentor or the faculty person can form groups or tell the students they must form their own groups to develop a particular program, and they must demonstrate who did what on it. They then present it as a group project online. Or in nursing, collaboration on a given research topic. Or in accounting, to do a financial statement from a real web site somewhere. For any given discipline area, there are ways of giving assignments that require this collaboration. And part of what makes these courses manageable is that the communication is mostly in threaded discussion areas which are archived and accessible throughout the term of the course (Carole Hayes, Coordinator, External Relations and Development, Office for Distributed and Distance Learning, Florida State University).

One administrator talked about requiring both individual and group projects.

In every lesson there will be at least one individual project and one group project (Joy Edwards, Director of Graduate Studies, Texas Wesleyan University).

Administrators and faculty organize groups in varying ways, sometimes forming the groups themselves, other times leaving it up to the students.

> I usually form the teams because people come in and don't know each other. I had four people come in and they want to be a team, so that team is already formed. But 90% come in and need help forming a team. I look at geographic areas and put them together (Joy Edwards, Director of Graduate Studies, Texas Wesleyan University).

At one institution students are assigned to teams.

> Students are also required to be part of a study team. This is not an independent study program. They are assigned to a team, most of the teams are face-to-face where there are other students in the same city that are also enrolled in the program. About one-third of the teams are online teams that don't meet face-to-face. Our university has a web page where chat rooms are set up and they go there to meet (Joy Edwards, Director of Graduate Studies, Texas Wesleyan University).

At other institutions, it happens in a more haphazard way.

> Mostly what happens is that students will find their fellow students nearby and get together on their own to meet (Carole Hayes, Coordinator, External Relations and Development, Office for Distributed and Distance Learning, Florida State University).

> Some have circulated email addresses, and students and/or faculty get together that way (Arthur Friedman, Professor and Coordinator, College of the Air, Nassau Community College).

> Yes, group work is a normal part. I write it all over my syllabus and elsewhere, what I call my promising practices to get a good grade. What I suggest is that they exchange phone numbers the first few classes, that they identify three or four people for their cluster group (cyber buddies, we call them). Share skills, exchange notes if you miss class, if someone has a tape recorder, share that. I've always done that, but now online they can have their own area for their group and not have to wait until next week to meet. It increases their trust level, and for those who are afraid of writing something and having twenty people look at it, this way only three look at it. This is the first time I've done that, and I still have a few

who have not used this area and I need to understand why not (Don Cardinal, Chapman University, School of Education Faculty member).

Some respondents commented on how the use of group projects works to bring those involved in a particular industry together so that students learn from the practical knowledge other participants possess.

> Students often have to do projects in their courses which involve working with other students, off-campus and on-campus. Which is wonderful for our graduate students here because they get to work with people in industry and get real insight from them. So that's one way.... It may not be required in every class, but it is typical for them to have projects (Elizabeth Spencer-Dawes, Manager, Distance Learning, Boston University).

> We also encourage students to develop projects as much as possible so that they have a product at the end. Hopefully the product relates to their job, and that makes it more interactive. A lot of times we will set these up as team projects. So the students interact with each other around the projects (Vice-President, anonymous large, independent, Eastern U.S. doctoral degree-granting institution).

> Because the delivery format can have some limitations, we felt that we needed something more than the traditional correspondence paradigm. The other reason is that most of the adult learning literature says that adult learners learn best with other adults, particular with those in the same profession, who can meet in a professional setting. And a lot of our teams meet in their schools. They meet after school hours in a classroom. So, we felt that this would almost require that there be some sort of professional collaboration. And in fact the subject matter requires professional collaboration (Joy Edwards, Director of Graduate Studies, Texas Wesleyan University).

Some respondents spoke about specific techniques used with teams in distance learning courses.

> But part of their grade is dependent on what they post to the threaded discussion areas that anyone can observe. So part of what a mentor does is observe what is going on in these groups, or an instructor can do this, and can come in at anytime and say, "Guys, this is looking real good. Let

me give you a clue. Move this way, or this is a good idea, or you need to follow through." That kind of interaction is always there. This provides an excellent teaching and learning tool plus a way of interacting with and managing large groups of students (Carole Hayes, Coordinator, External Relations and Development, Office for Distributed and Distance Learning, Florida State University).

Phone conferences with the team and the mentor are required, at least two a semester. So the entire team has to get on the phone (Joy Edwards, Director of Graduate Studies, Texas Wesleyan University).

Some administrators spoke about the techniques for using threaded discussion areas and chat.

We know groups are effective, but you really learn a lot when you can read every word of a group discussion. It's not just by listening in to a few words and making assumptions, which are truly assumptions, but it is based on fact. I have areas for mid-term discussion, I have an on-going one for evaluating the class. In the first week someone always writes, I don't think Don understands, I'm just not good at math. I'm going to have to take another class. Right about now they aren't happy with their grades, but I can say look at what you've learned (Don Cardinal, Chapman University, School of Education Faculty member).

We have an instructor that uses threaded discussed areas for peer review on essays (Greg Chamberlain, Dean of Learning Resources, Bakersfield College).

Group work in the sciences was not discussed by the administrators interviewed for this study. Perhaps this is because of the greater difficulty for students to carry on discussion of this subject matter in online platforms.

Some respondents spoke about the variety of approaches to developing a sense of community in distance learning courses, mostly as a result of leaving this issue up to faculty members.

That's up to the instructor. We encourage instructors to use collaboration, team projects, and group discussions, but every instructor handles online classroom activities differently. We don't dictate how they run their online classes (Vivian Sinou, Dean, Distance & Mediated Learning, Foothill College).

The instructors are all over the map. Some use chat rooms, more of them use threaded discussions that have specific objects that they are graded on (Thornton Perry, Director of Distance Education, Bellevue Community College).

In summary, the interviews with distance learning administers reveal that teams are used extensively and can be effective in creating a sense of community. Administrators spoke about various specific techniques such as group projects, ways of forming groups, and the use of chat rooms and threaded discussion areas.

TO GROUP OR NOT TO GROUP

The research literature shows advantages for students working in groups, particularly in online environments. It helps motivate students, builds a sense of community, and increases learning. However, as indicated in the previous chapter, often independent-minded students drawn to distance learning do not want to interact with other students. The 2001 survey found that 73.3% of computer-based course students responded that they either disagreed or strongly disagreed with the statement, "I like to have group projects and other opportunities to learn in group situations."

Notice here that only half (50%) of those in videotape courses with no group interaction responded that they disagreed or strongly disagreed with the statement.

In terms of interdependent tasks, respondents once again to the 2001 survey indicated a surprising lack of interest in taking part in this type of group learning. In this survey, 63.2% of the computer-based students either disagreed or strongly disagreed with the statement, "I would prefer to communicate with other students around interdependent tasks, rather than open socializing opportunities at a distance."

Figure 9: Prefer Group Learning (Questions 1 & 10)

Delivery format * I like to have group projects and other opportunities to learn in group situations. Crosstabulation

			I like to have group projects and other opportunities to learn in group situations.				Total
			strongly agree	agree	disagree	strongly disagree	
Delivery format	computer-based	Count	2	14	29	15	60
		% within Delivery format	3.3%	23.3%	48.3%	25.0%	100.0%
	videotape	Count	4	27	25	6	62
		% within Delivery format	6.5%	43.5%	40.3%	9.7%	100.0%
	correspondence	Count	1	3	1		5
		% within Delivery format	20.0%	60.0%	20.0%		100.0%
	other	Count	1				1
		% within Delivery format	100.0%				100.0%
Total		Count	8	44	55	21	128
		% within Delivery format	6.3%	34.4%	43.0%	16.4%	100.0%

Figure 10: Prefer to Communicate with Other Students Around Interdependent Tasks (Questions 1 & 11)

Delivery format * I would prefer to communicate with other students around interdependent tasks, rather than open socializing opportunities at a distance. Crosstabulation

			I would prefer to communicate with other students around interdependent tasks, rather than open socializing opportunities at a distance.				
			strongly agree	agree	disagree	strongly disagree	Total
Delivery format	computer-based	Count		21	27	9	57
		% within Delivery format		36.8%	47.4%	15.8%	100.0%
	videotape	Count	3	30	24	3	60
		% within Delivery format	5.0%	50.0%	40.0%	5.0%	100.0%
	correspondence	Count		1	2		3
		% within Delivery format		33.3%	66.7%		100.0%
	other	Count		1			1
		% within Delivery format		100.0%			100.0%
Total		Count	3	53	53	12	121
		% within Delivery format	2.5%	43.8%	43.8%	9.9%	100.0%

Again, the students in videotape courses with no interaction with other students disagreed at a lower rate (40.7% "disagree," 5.0% "strongly disagree") with the statement. Clearly, there is a somewhat puzzling lack of interest to work in teams by some of the respondents. In looking at the issue of group work in online environments, one needs to be conscious of the fact that many of the more independent-minded students drawn to this delivery method resist teams.

TRADITIONAL NOTIONS OF COMMUNITY IN HIGHER EDUCATION

While it is a commonplace criticism of distance learning to claim that there is a lack of community in computer environments, the notion of community in higher education has become increasingly difficult to accurately define. The modern university is a complex institution, a collection of distinct departments, student groups, administrators, government agencies, and faculty members. This complex, modern university has been described as a "multiversity" by Clark Kerr (1963) to distinguish it from its historical predecessor that was thought to be a unified community of masters and students with a single vision—the quest for truth. Kerr's sense of the modern university is one of a pluralistic institution with no single community: "...what had once been a community was now more like an environment—more like a city, a 'city of infinite variety'" (Kerr, 1963, p. 102). Nevertheless, in recent times many have criticized traditional higher education for lacking a stronger sense of community (Abelson, 1997). Furthermore, it is reported that students and faculty are longing for a deeper sense of community (McCarter,

1996). Taken altogether, it is clear that community is not just a problem with distance learning, but also something with which traditional higher education is struggling at present. So as we proceed here to delve deeper into the group method, it is important to understand the broader issue of community in higher education.

SOCIAL OR GROUP LEARNING THEORIES

In traditional educational environments, there has been a growing organization of learning in groups with an increased use of teams and group projects. We saw in Chapter One that popular learning theories of Vygotsky focused on social factors in learning. How do social factors contribute to learning? Goldman (1999) claims that traditionally education is seen as an activity of isolated thinkers pursuing truth in a spirit of American self-reliance. However, in practice, education is very much a social activity, especially the research component that is heavily dependent on colleagues. In fact, some argue that the key to the learning process as a whole is the interaction among students, and between faculty and students (Palloff & Pratt, 1999).

Theories of the importance of social aspects to learning have become increasingly fashionable in the educational theory literature. Spector (1999) notes that this social learning theory perspective draws heavily on Bruner, Lave, Piaget, and Vygotsky. Two of the most discussed current approaches to learning in teams are cooperative and collaborative learning. Each represents opposing ends of constructivist teaching-learning, ranging from highly structured by the teacher (cooperative), to one that gives the responsibility for learning primarily to the student (collaborative).

Cooperation is a structure of interaction designed to facilitate the accomplishment of a specific end product or goal through people working together in groups. Cooperative learning is defined by a set of processes to help people interact together in order to accomplish a specific goal or develop an end product, usually content specific. It is more directive than a collaborative system and closely controlled by the teacher. While there are many mechanisms for group analysis and introspection, the fundamental approach is teacher centered whereas collaborative learning is student centered. Cooperative learning is based on the creation of systematic application of structures or content-free ways of organizing social interaction in the classroom. An important aspect of the approach is the distinction between "structures" and "activities." In terms of student motivation, social theory assumes that cooperative efforts are based on intrinsic motivation generated by a joint aspiration to achieve personally significant goals. Contrary to this, behavioral learning theory assumes that cooperative efforts are powered by extrinsic motivation to achieve rewards.

Collaboration is a philosophy of interaction where individuals are responsible for their actions, including learning, and respect the abilities and contributions of their peers. Collaborative learning is a personal philosophy, not just a classroom technique. In all situations where people come together in groups, it suggests a way of dealing with people that respects and highlights individual group members' abilities and contributions. In this approach there is a sharing of authority and acceptance of responsibility among group members for the group's actions. The underlying premise of collaborative learning is based upon consensus building through cooperation by group members (Bruffee, 1995; Panitz & Panitz, 1998).

Ken Bruffee (1995) argues that what determines which approach is used depends upon the level of sophistication of the students involved, with collaborative learning requiring more advanced student preparation than cooperative learning. He identifies two types of knowledge as a basis for choosing an approach: foundational and non-foundational. Foundational knowledge is the basic social knowledge generally agreed upon, such as spelling and grammar, mathematics, and historical facts. Bruffee claims that this foundational knowledge is best learned using cooperative learning structures in grade school. Non-foundational knowledge is attained through reasoning and questioning rather than rote memory. The other way non-foundational education differs from foundational is that it encourages students not to take their teacher's authority for granted. According to Bruffee, collaborative learning shifts the responsibility for learning away from the teacher as expert to the student. Using this model, most adult learning lends itself to a collaborative approach.

Bruffee (1995) sees education as an acculturation process occurring through conversation. Students learn about society by developing the appropriate vocabulary and by exploring norms in conversation. He views the two approaches as connected, with collaborative learning designed to pick up where cooperative learning leaves off. In effect, students learn basic information and processes for interacting socially in the primary grades and then extend their critical thinking and reasoning skills and understanding of social interactions as they become more involved and take control of the learning process through collaborative activities. This transition may be viewed as a continuum from a closely controlled, teacher-centered system to a student-centered system where the teacher and students share authority and control of learning.

Although these represent two different approaches, many of the elements of cooperative learning may be used in collaborative situations. If adult learners work in collaborative learning environments, then they must have an understanding of how to work with others and value individual contributions. This suggests that computer software should be designed to support such an understanding and appreciation of learning in groups. Nevertheless, although the research literature on the benefits of students working in groups is deep, Schwartz (1999) claims that over sixty years of research has not shown that the work of the group is stronger than the most capable individual member.

RESEARCH LITERATURE ON GROUP METHOD IN COMPUTER ENVIRONMENTS

Littleton and Hakkinen (1999) argue that computers can form a particularly rich context for understanding collaborative learning and may assist researchers in better understanding the benefits of learning in groups. Furthermore, computers may lead to a better understanding of human ability to collaborate in learning environments. Computers may provide a mechanism to handle the awkwardness of group work and clarify the importance of representing participant thoughts to others during collaboration.

King (1998) notes that numerous scholars have analyzed the communication occurring in threaded discussions and chat rooms for a better understanding of educational approaches. Analysis has included socio-cultural perspectives, levels of interactivity, emergent comment categories, and participant roles. Surprisingly, Kirkley, Savery and Grabner-Hagen (1998) found that most email interaction was from individuals to groups (19.0% instructor to whole class; 68.9% student to whole class). Student-to-instructor and student-to-student interaction amounted to only 3.1% of the communication each. This is a curious data set. Perhaps students are happy with more generalized communication to the group as a whole and don't feel the need for one-to-one communication. In traditional courses, Nunn (1996) found that in a group of twenty university courses in a public institution with enrollment between fifteen and forty-four students, only two percent of the time was taken up in discussion, and most of that was directed by the faculty member rather than being student-centered. One can argue about the precision of the data, but there appears to a difference in the communication that occurs in distance learning courses when compared to the traditional classroom. Rourke, Anderson, Garrison and Archer (1999) describe this difference in a notion of "social presence" in online courses, defined as the ability of learners to project themselves socially into a community of inquiry.

Dillenbourg (1999) argues that "collaborative learning" in distance learning is neither a mechanism nor a method. It is a situation whereby specific forms of communication are expected to occur and lead to learning. However, these communications are complex and difficult to understand. He sees the challenge as to understand and control interaction in collaborative learning situations. Additionally, asynchronous communication makes understanding the communication even more difficult, adding another wrinkle. Dillenbourg views collaborative learning as dependent on a shared conception of a problem. In collaborative environments, it is important to define the situation, interactions, processes, and effects.

According to Duffy, Dueber and Hawley (1998), students' problem-solving abilities can be judged by evaluating and distinguishing significant from insignificant information. This ability of the students is important to understand because, as we saw earlier, there are indications that this sorting process of the content is a key instructor role, one that can also occur through interaction with a group of learners. The higher the degree of problem-solving ability in students, the less they need faculty members to assist them in this prioritizing function.

Grounding is a key concept repeatedly mentioned in the research literature. "Grounding" is defined as mutual understanding about problems and tasks for groups of learners (Baker, Hanse, Joiner, & Traum, 1999). A high degree of grounding is necessary for collaborative learning. Semantic-level grounding has to do with understanding the meaning of particular words and symbols, whereas practical grounding focuses on what needs to be done. An understanding of agents, tools, and goals is essential for grounding. Collaborative learning requires participants to increase effort towards understanding shared meanings and semiotic tools for effective interaction.

CHAT is a theory of group learning that is also referenced often in the research literature. Cultural-Historical psychology/Activity Theory (CHAT) is an integration of various interdependent relations in the learning process, a dialectical process (Hansen, Dirckinck-Holmfeld, Lewis & Rugelj, 1999). Derived from Vygotsky's cultural-historical school of psychology that led to "activity theory," this theory points out the importance of tools to mediate knowledge acquisition. As all learning activity is in some way mediated by tools, and tools are developed by the larger social world, this social sense is involved in all learning. Students are collaborators through the tool. A central notion of the activity theory is that there are three levels: intentional, functional, and operational.

Spector (1999) argues that supporting active participation and reflection is a key to success in group learning. Computer-Supported Collaborative Learning (CSCL) is a notion focused on supporting this activity by creating shared communities of practice. Spector recommends that the following basic principles are characteristic of CSCL: support of collaborative analysis of problems, providing tools for collaboration, and facilitating divisions of labor. Similarly, Chiu, Chen, Wei and Hu (1999) recommend that effective cooperative learning with computers include group structure, task structure, incentive structure, individual accountability, and cooperative space. The need is for software to support these cooperative conditions.

The research literature also includes a variety of specialized topics. According to Joiner, Issroff and Demiris (1999), methods for understanding robot-to-robot communication can help with utilizing human-to-human communication in learning environments. They claim that research has shown mixed results in looking at the role of talking in group projects. Additionally, they claim that certain kinds of tasks lend themselves more to group work. Concentrating on interpersonal issues, Savery (1998) argues that more successful teams are aware of each other's lives outside the classroom and have personal connections. Also, the presence of clear group leaders leads to more successful teams.

One can see from this brief review that much remains to be done in researching group interactions in online educational environments. However, the indications are that the communication occurring in online settings is both different and more visible than in the traditional classroom. The challenge for software designers is to understand how these two characteristics can be used to the advantage of learners.

VIRTUAL TEAMS IN BUSINESS

Over the last few years there has been a great deal of discussion about the use of teams in the business setting. Much of the research literature is relevant to distance learning, particularly in its use for the adult learner. As companies have become increasingly multinational and distributed geographically, the formation of virtual teams has been a natural occurrence. Research on virtual teams shows that issues of interdependence, leadership, social communication, and project management are key in forming successful teams (Lipnack & Stamps, 1997; Jarvenpaa, Knoll & Leidner, 1998). A primary advantage detailed in the research literature is that connections are made through the sharing of ideas and thoughts. How people look and their cultural, ethnic, and social background are becoming irrelevant in this medium—consequently some refer to it as the great equalizer. Furthermore, introverts tend often to be more adept at using a virtual rather than a face-to-face environment.

Communities are formed around issues of identity and shared values, according to Palloff and Pratt (1999). Palloff and Pratt define the principles of building community online as: defining the purpose of the group, creating a distinctive gathering place for the group, promoting effective leadership from within, defining norms and a clear code of conduct, allowing for a range of member roles, facilitating subgroups, and having participants resolve their own disputes. They see virtual versus human contact as a false dichotomy because people can communicate both ways. They argue that the failing of many computer-mediated distance learning programs has been due to the inability to facilitate a collaborative learning process.

What is sought is teamwork and collaboration aided by technological tools for communication, information retrieval, and simulation.

INTERDEPENDENT TASKS

Interdependent responsibility for competing tasks is a major aspect of virtual teams used in business. The basic elements of virtual teams according to Lipnack and Stamps (1997) are people, purpose, and links. Virtual teams are groups of people working across space and time who interact by webs of communication technologies through interdependent tasks guided by common purpose. Cook (1995) argues in a similar way that shared interests build community in distance learning. The method of small group instruction used in traditional courses needs to be used to build community in distance learning. Additionally, because of the lack of in-person contact, students need more regular and consistent feedback from the instructor. Luetkehans and Bailey (1999) argue that teams are affected by the type of action, task, and the nature of the problem. They classify activities as motivational, problem-solving, skill building, and knowledge construction. Additionally, they argue that large teams are less productive in online environments and that team size should be limited to three or four persons.

The focus on interdependent tasks in business teams is one that may be particularly useful in educational environments. This idea fits well with project-based education and collaborative learning methods. Furthermore, adult learning theory points to the importance of integrating learning with work, making this approach useful for distance learning. Finally, the indication that group projects can increase a sense of community is exciting and may address the problem cited so often by educators in the field.

LEADERSHIP

Leadership takes on a very different form in virtual teams. In fact this is one of the main challenges in the formation of virtual teams. In virtual teams, leadership based on hierarchy and job title is not effective (Lipnack & Stamps, 1997). Instead, leaders are recognized based on their skills and communication abilities. Furthermore, successful teams often have a system of multiple leaders who seem to rotate in taking the lead. Indeed, high performance teams are characterized by rotating leadership (Jarvenpaa, Knoll & Leidner, 1998).

In one way the democratization of leadership in online environments as opposed to face-to-face is an obvious advantage in motivating reluctant leaders, but also may lead to more role confusion, or a leadership vacuum. Assigning students the responsibility for leading a portion of the discussion online is one strategy supported by Palloff and Pratt (1999). Active student roles might include facilitator of discussion, process observer, group learning summarizing, team leader, and presenter.

SOCIAL COMMUNICATION

We saw earlier how respondents to the 2000 survey used various technologies to encourage socializing among the students. However, despite common approaches and perhaps common classroom wisdom, free socializing may be ineffective. Instead, focus on interdependent tasks may be the key. One wonders here if the nature of the self-selected students in these courses give them as a group more of a no-nonsense socializing attitude? In one study, Jarvenpaa, Knoll and Leidner (1998) found that effective teams were task oriented and rarely engaged in social comments. Furthermore, this study found that communication centered on assignments and very rarely took a broader social form.

Evidently, in this study the importance of trust became based on group member ability to work towards the common group goal, not on personal bonds developed in a more general socializing environment.

PROJECT MANAGEMENT

One problem virtual teams experience is difficulty in managing a project as it proceeds with team members in geographically dispersed areas completing different tasks (Lipnack & Stamps, 1997). The problem is that often valuable information about past project history is not captured or retained or maintained in a useable format. Tools are needed to help with this capturing of the thinking thread of the project (Weiser & Morrison, 1998). We will see later in the conclusion how we might begin to create these tools and what their necessary attributes would be.

THREADED CONVERSATIONS/CHAT

When considering social communication in online environments one must think about the use of chat versus asynchronous threaded discussions. What are the advantages and disadvantages of each? Jonassen (2000) argues that asynchronous conversations are usually more reflective and thought out than synchronous ones, while it is easy to lose focus in synchronous conversations. He points out that asynchronous environments have the advantages of allowing students to argue without excessive conflict, providing a safe format where introverted students can excel.

Jonassen points out the more obvious disadvantages of online dialogues in their reliance on text communication and absence of visual cues from students. Personal experience with chat indicates that it is often ineffective and best used for small work groups. The disadvantage of threaded discussions is that they are overly structured, appearing often as a series of short speeches instead of communication. How might this be improved?

COMPUTER-MEDIATED CONFERENCING

One solution is what Gilly Salmon (2000) calls "Computer-Mediated Conferencing" (CMC). She defines a five-step model for stages of student development in CMC including motivation, socialization, information exchange, knowledge construction, and development. Salmon argues that CMC is defined by the lack of traditional hierarchies, an ability to synthesize knowledge, and different styles of communicating. She claims that in the future the unbundling of the components of teaching will lead to CMC moderating as a discrete function. For Salmon, understanding the social aspects of learning is more apparent in online environments. This is because distance learning is about the facilitation of communication—not posting content. Models that see community formation as a structured process while accommodating individual differences are central here. In agreement with what was indicated earlier in this chapter, Salmon points out that many of the benefits of CMC come from building an online community that is working together at common tasks. She holds that education and training are always undertaken for a purpose. Unlike casual browsing or playing computer games, a key distinction of online education from other online activities is that it is very purposeful. Salmon argues that online courses need accepted ends to work towards, and the means to get there. She sees a trend away from courses towards supporting professional development and performance. Knowledge acquisition will be richer, broader, and linked to personal experience in complex ways.

At the British Open University, most courses include some face-to-face sessions and do not yet include CMC. She claims that the lack of face-to-face information about participants encourages real exchanges. Additionally, the text basis of CMC means that conversations are more permanent. Salmon points out that participants are better able to follow and contribute to one dominant thread than many at the same time. Furthermore, in threaded discussions students should try to write postcards, not letters, a model that looks for straightforward and short communication. Additionally, as we have seen others emphasize, in order for CMC to fulfill its promise, it needs to figure out how to make use of previous online discussion.

According to Salmon, tutors are a key in the British Open University system, and many ascribe the success of this system to the tutor system. One should note here that the British Open University tutorial system accommodates optional communication and learning through groups. CMC is now changing the face-to-face system of the British Open University, which previously had a tutor working with up to 25 students mainly through a postal system and group tutorials at local study centers. She sees the role of e-moderators as pulling together participants' contributions and relating them to concepts and theories that are the subject of the course. Salmon argues that computer-mediated conferencing actually requires e-moderators to have a wider range of expertise than the traditional classroom. Online learning offers participants opportunities to explore information rather than asking them to accept what the teacher determines should be learned. Salmon claims that educational institutions of the future will be most successful if they encourage and develop skilled online moderators.

Finally, Salmon notes that most CMC users are looking for pragmatic and realistic approaches to learning and have little patience for obscure, academic methods. As individuals become more networked, they want smaller chunks of relevant knowledge. In terms of learning theory characteristics, Salmon argues that they look for rules and guides for understanding the course organization early in the process. Furthermore, students are now accustomed to computer environments and want immediate, interactive, and visually based learning. Additionally, women's ways of communicating and working may lend themselves better to CMC than men's, which tends to be more individualistic.

CONCLUSION

The etymology of community is the Latin "communitas" meaning common and is defined as a body of people associated by common status and pursuits (Hoad, 1986). The notion of community associated by pursuits is one that is useful in relationship to education—it leads to rich connections with both constructivist learning theory and lessons learned from teams in business environments. Kerr's notion of the higher education community as an environment or a city is one that is evocative in relationship to distance learning because the community becomes a total environment, not just a series of relationships. These ideas about community taken together lead us to an understanding of community in higher education as a complex environment built on common purpose and interdependent pursuits.

What can higher education learn from the experience of virtual teams? Virtual teams parallel learning communities in many ways: they meet the needs of geographic and time convenience, they are usually formed temporarily, and involve members with often vastly different backgrounds and skills. However, the primary likeness is that they facilitate communication among members who would not ordinarily be able to form teams because of time and place constraints (Lipnack & Stamps, 1997; Jarvenpaa, Knoll & Leidner, 1998). How are virtual teams different from current distance learning programs? The two key differences are that often groups of students are not seen as teams and that they are not project-based. Virtual teams used in an educational setting can increase the sense of community through purposeful activity. Furthermore, the role of the instructor in distance learning courses is problematic. Many have written about the shift of the role of the instructor to that of more of a facilitator or course designer (Daniel, 1997; Davis & Botkin, 1994), and this is confirmed in the experience of virtual teams.

In addition, often in current distance learning courses, there is an attempt to build community through inserting socializing opportunities. This effort takes the form of introductions, chat rooms and discussion groups set up outside the formal context of the course. What the experience with virtual teams reveals is that these artificial attempts to create community would be more effective if structured within the context of the group project and related tasks. In fact, explicit socializing may not be necessary for the formation of community, but simply grow naturally out of effectively formed teams working towards a common purpose.

In response to the findings of this literature review of community in higher education, current trends in higher education, and virtual teams, I propose a model called a "virtual learning community" to address the problem of community in distance learning. A virtual learning community is defined as a group of learners that is networked with other learners, knowledge media, and a facilitator all working towards the common purpose of acquiring knowledge through interdependent pursuits. Figure 11 expresses the clustering of virtual learning community members, methods, and learning technologies and indicates some of the connections. The subject at the center of the figure constitutes a viewpoint that is a shift from either a student- or faculty-centric model of higher education. The outer ring of the circle shows the major participants in the learning network: other learners in a course, faculty, work, and the knowledge media. Figure 11 shows how each particular relationship in the learning community is mediated by a particular method and technology. (Although "knowledge media," the phrase Open University tends to use, has a broader meaning that encompasses traditional notions of content, method, and technology.) One can read the figure starting from the "work" box and move inward to see that they are enabled by simulations, tools to integrate learning with work, and on-going learning tools. Moving further into the methods circle, we see that they are then supported by notions of adult learning theory, assessment of student needs, and reality-based knowledge. Of course, many of the individual elements in this figure have multiple connections.

How are virtual learning communities different from traditional higher education models? First, a virtual learning community begins by substituting student with learner in order to focus the structure on the needs of the learner instead of only the demands of the teacher. The most important aspects of a virtual learning community are the concentrating of the process on the learner, the increased contribution of both knowledge media and other learners, and the importance of integrating learning with the workplace. Second, virtual learning communities need to recognize the importance of knowledge media as one of the primary nodes in the distributed learning community. Knowledge media is a concept developed by the British Open University (Eisenstadt, 1995) to define the convergence of computing, telecommunications, and the cognitive sciences. It represents a change in the relationship between people and knowledge in a fundamental way (Daniel, 1997). Knowledge media would generally not be thought of as part of the traditional university community because it is seen as static information or content, but it is in fact dynamic, interactive knowledge. While some may argue that knowledge media should not have such large importance in the learning community, it has been a key member of the academy since the beginning in the form of text. The majority of a student's learning at the university is done through the use of a single medium (text), and this is formalized by accrediting agencies in the requirement of approximately two hours outside the class (mostly involving reading and writing) for each hour of seat time. The combination of computing and telecommunications make it more evident that knowledge media is more than just content. This is a key point to grasp in the transition to virtual learning communities.

As the importance of knowledge media is increased in the learning community, the primary role of the instructor is shifted. In terms of economic models, distance learning attempts to do this in a very conscious manner by transforming the large labor cost of face-to-face teachers into the accumulation of knowledge media capital (Daniel, 1997). As John Daniel, the head of the British Open University, points out, there are two economic models for distance learning: one which seeks to augment traditional classroom courses, and one which seeks to replace. The augmenting model leads to increasing the cost of education, while the replacement model leads to the reduction of costs through the investment in knowledge media capital instead of ongoing labor expenses. See "Why Distance Learning?" (Berg, 2002) for more on this subject.

The most important conclusion of this chapter is that community in distance learning courses can be created through team formation. While every distance learning course on every subject may not be possible, an understanding of this principle can go a long way towards creating a sense of community. Furthermore, by clearly defining the nature of educational communities as being based on common purpose, and by examining the research on virtual teams, it is clear that virtual learning teams should be structured around common projects and interdependent tasks.

Clearly, much can be learned from virtual teams in relationship to distance learning in higher education. First, in seeking to form a sense of community, project-based curriculum can be extremely useful. Secondly, many of the traditional ways of attempting to form community in distance learning courses, such as creating opportunities for personal communication, are more effective when contextualized within interdependent projects. Third, the role of the instructor is vastly altered and leadership must be given up to a certain degree to other members of the team. Fourth, the educational software needed for distance learning will need to assist with communication and with overall project management. In looking forward to the primary characteristics of devices that will assist learners in virtual learning communities, enabling communication and the formation of virtual learning communities will be primary. In addition, project management and the recording and exchange of thoughts must be part of new learning devices.

Finally, networked learners can actually increase a sense of community to an extent to which is just beginning to be realized and may even work to address the complaints of lack of community in traditional higher education. Although there has been much public discussion and paranoia about virtual communities and distance learning, some have pointed out that this is not an either/or proposition (Etzioni & Etzioni, 1997). What needs to be done is to identify the opportunity that virtual communities represent. Michael Dertouzos (1997) points out that what he calls the information marketplace will lead to great advances in education because it represents the first technology advance that directly links to the acquisition, organization, and transmission of knowledge and the simulation of processes at the heart of the educational process. In spite of the inadequacies of many of the distance learning programs created to date, it is inevitable that technology will be a key in the transformation of higher education and that virtual learning communities and the use of teams will prove to be a missing link in addressing the important social aspect of the learning process.

In the next chapter, we turn to the extensive research on human-computer interaction as it applies to computer educational environments including human factors, usability, interface design, hypertext, metaphor, and task analysis. In looking at this extensive and sometimes technical literature, it is useful to keep in mind the theories presented in the first three chapters on learning theory, as well as individual and group approaches to learning.

Figure 11: Virtual Learning Community

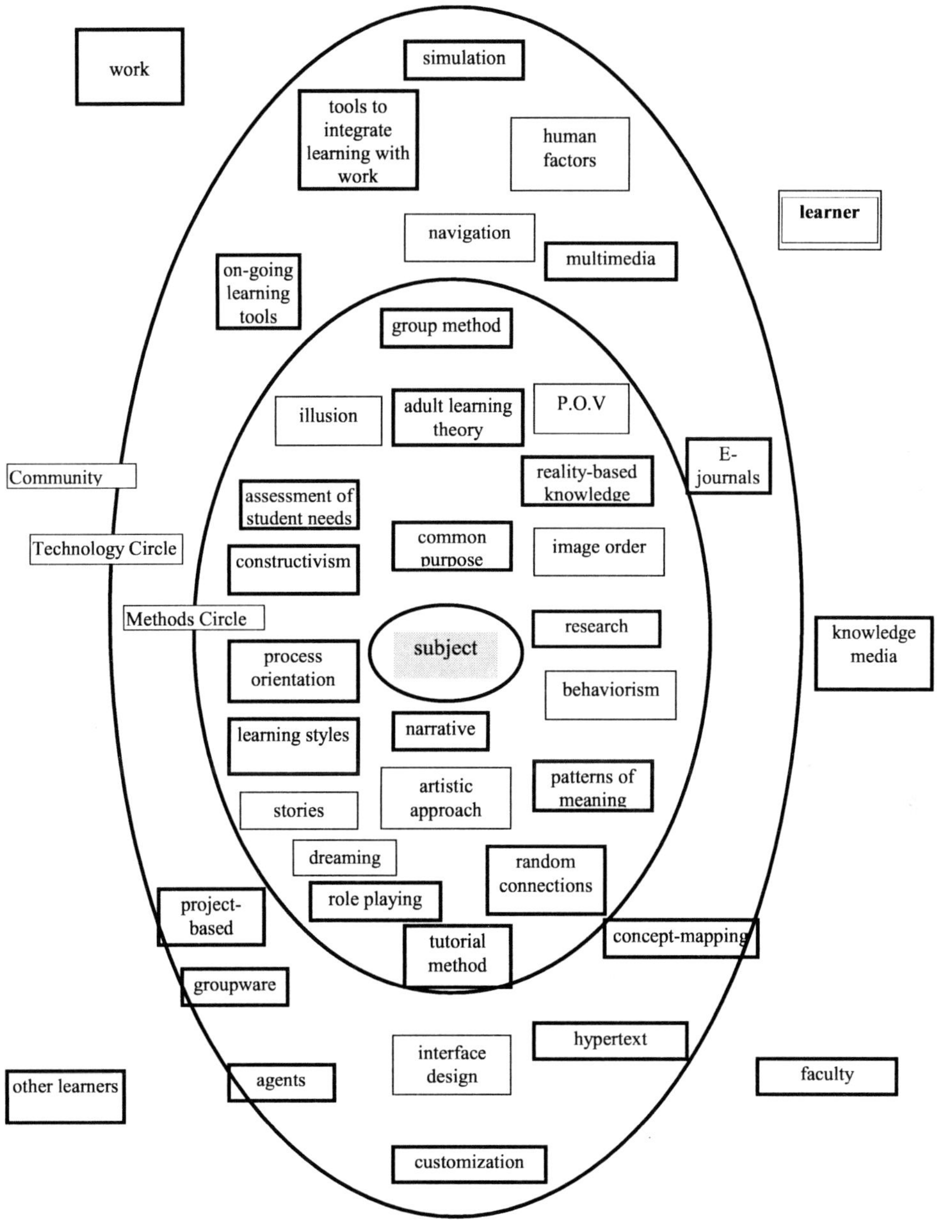

Chapter V

Human-Computer Interaction in Education

This chapter seeks to understand aspects of Human-Computer Interaction (HCI) research relevant to education. The general study of human-machine interaction began in World War II with a focus on understanding the psychology of soldiers interacting with weapon and information systems such as signal detection and cockpit instrument displays. After the war, human-machine interaction began to be examined more broadly in relationship to work and consumer product environments. Human-computer interaction developed from this work and is a multi-disciplinary field involving computer science, psychology, engineering, ergonomics, sociology, anthropology, philosophy, and design. HCI is concerned with the design, evaluation, and implementation of interactive computing systems for human use (Card, Moran & Newell, 1983; Faulkner, 1998; Head, 1999; Helander, 1998).

The subject of HCI has been assigned various labels and acronyms over the years. The acronym is generally used to mean human-computer interaction, but sometimes is described as human-computer interface. Additionally, CHI, or computer-human interaction, is sometimes used, as well as MMI, man-machine interface. The primary focus of HCI is the user—the field as a whole tries to better understand the interactions between the user and computer. The primary factors considered in examining human-computer interactions are organizational, environmental, cognitive, task, constraints, and functionality (Faulkner, 1998; Head, 1999; Maddix, 1990; Preece & Shneiderman, 1995).

Cognitive research and principles developed in the 1980s provided much of the early HCI framework (Faulkner, 1998). The literature on HCI focuses in part on cognitive processes, especially in terms of the capacities of users and how these affect users' ability to carry out specific tasks with computer systems. In contrast to behaviorism, which argues that action must be understood in terms of observable behavior between humans and the environment, cognitive psychology focuses on mental processes sometimes expressed in computational terms (Wooffitt, Fraser, Gilbert & McGlashan, 1997). In terms of cognitive issues, HCI concentrates on motor, perceptual, and cognitive systems and two types of memory: working and long-term. According to Card, Moran and Newell (1983), the most effective technique for retaining information is to associate it with something already in long-term memory. Thus, much of this literature on the cognitive aspects of HCI is concerned with the relationship between long- and short-term memory. One might draw parallels here to Piaget's learning theories and concepts of accommodation and assimilation. Accordingly, memory is broken down into the following aspects: processor cycle time, memory capacity, memory decay rate, and memory code type. Obviously, these cognitive issues have special importance for educational applications. How can computer applications best make use of memory attributes to increase learning?

HUMAN FACTORS

HCI is a subset of the field of human factors that also includes interface design, system/user communications, and end-user involvement (Carey, 1991; Reisner, 1987). Human factors as a field is defined by Carey (1991) as "the study of the interaction between people, computers, and their work environment" (p. 2). The objective of human factors research is to create information systems and work environments that help to make people more productive and more satisfied with their work life. However, the overall emphasis of human factors is on system performance, not on human satisfaction. Today, most computer and software companies have human factors staff (Helander, 1998), and Shneiderman (1987) claims that the diverse use of computers is stimulating widespread interest in human factors issues. He points to five primary human factors: time to learn, speed of performance, rate of errors by users, subjective satisfaction, and retention over time. All of these factors are central to learning—and here one can see how closely learning theory parallels human factors research.

Human error studies are part of the human factors literature. This research has taken two different routes: natural science and cognitive science approaches (Reason, 1990). HCI is concerned with the cognitive science approach and we shall see that the field is very much focused on learning how to minimize user error. Reason (1990) identifies three basic kinds of errors: skill-based, rule-based, and knowledge-based. He argues that errors are bound with stored knowledge structures retrieved in response to situational demands. While this research on human error is concerned primarily with preventing computer users from having trouble using the software, it is vital in assessing student work that incorrect responses are a result of user performance, not software design.

USABILITY

Another major area of study that overlaps with HCI is usability. Usability refers to how a computer system is effectively manipulated by users in the performance of tasks (Carey, 1991). Usability evaluates if a computer system functions in the manner it was designed—if it fits the design purpose (Faulkner, 1998). This evaluation of usability includes the user interface, dialogue design, cognitive match with the user, quality of documentation, and online help. Interface design is one aspect of usability (Cohen, 1997). As opposed to the traditional mechanical point of view, usability focuses on the cognitive and social aspects of users when designing computer applications as well. Consequently, usability has a communications medium dimension in mediating between users and the designer. In this way, usability focuses on the evolving process of communication and supporting organizational processes (Adler & Winograd, 1992). Maddix (1990) emphasizes the process aspect of usability by suggesting a parallel with the concept of "gestalt," implying the understanding of computer systems as a totality, rather than as a collection of individual parts.

Shneiderman (1999) argues that designers of older technologies such as telephones and television have reached the goal of universal usability, but computers are still too difficult to use. Designing for experienced users is difficult, but designing for a broad audience of unskilled users presents a far greater challenge. Consequently, Shneiderman suggests three usability principles: supporting a broad range of hardware and software, accommodating users with different skills and needs, and bridging the gap between what users know and what they need to know. The principle of accommodating users with different skills and needs is particularly important in educational applications.

The literature on usability also includes information on access for special needs populations. The ACM's Special Interest Group on Computers and the Physically

Handicapped (SIGCAPH) promotes accessibility for disabled users. The European conferences on User Interfaces for All also deals with interface design strategies, and the Web Accessibility Initiative of the World-Wide Web Consortium has a guidelines document to support special needs users (http://www.w3.org/WAI). For educational institutions with mandates to provide access to various populations, accommodating students with special needs is crucial.

INTERFACE DESIGN

Computer interface design is a subset of HCI and focuses specifically on the computer input and output devices such as the screen, keyboard, and mouse. Research on the task interface has its roots in the ergonomic study of instrument panels during World War II. This research has led directly to the current computer interface design literature (Sime & Coombs, 1983). In a general way, interface design is the subject of this book. However, most of the interface design literature focuses broadly on computer software design, not just educational applications. Much of this literature focuses on principles of good computer interface design.

The best work in the field addresses characteristics of the users and conceptual issues of use to educational software designers. Donald Norman (1998, 1988, 1987), one of the leading researchers in this field, suggests seven principles of good design: using knowledge of both the world and in the user's head, simplifying the structure of tasks, making functions visible, using conceptual maps, exploiting constraints and limitations, expecting user error, and standardizing functions. Head (1999) references IBM's four design principles in recommending a focus on users, continual user testing, interactive design, and integrated design. Additionally, the literature is full of design truisms that tend to be repeated, such as consistency eases learning and use no more than four colors (Head, 1999). Oddly, these design "tips" are found both in the academic-style literature and the more popular design guides.

GOMS MODELS

I now turn to the primary areas of debate and research within HCI. The GOMS model of Goals, Operators, Methods, and Selection is one of the basic HCI principles often discussed in the literature (Card, Moran & Newell, 1983). Goals are set to provide a memory point for return if there is failure or to refer to for a navigational history. The operator is output such as the keyboard and mouse. Methods are learned procedures that the user already knows, rather than plans created during the completion of a task. Selection is the use of a set of selection rules, often using an if-then logic. GOMS was one of the first attempts to infer a cognitive model to describe how users perform tasks, and some see GOMS as a major advance in looking at models that predict human behavior (Reisner, 1987). Black, Kay, and Soloway (1987) see GOMS as a well-developed model for the study of story and narrative understanding in computer environments, particularly when utilizing the if-then logic that is similar to the standard plot device of question-answer (see Chapter Eleven for more on this). Recently more complex models have been proposed, using variations on linguistic grammar theory and production systems, and task-action grammar (Wooffitt, Fraser, Gilbert & McGlashan, 1997).

Wooffitt, Fraser, Gilbert and McGlashan (1997) criticize the GOMS method because users often behave by first acting when thrown into a situation, and only then devising a goal afterwards. They argue that an individual's actions are produced on a moment-by-moment basis, and that behavior in particular circumstances is not rule-governed. Consequently, the GOMS method may have limitations. As it is common in educational software to set clear learning objectives and goals, it is interesting to think about what happens when goals are negotiated after users are thrown into a computer environment. One should ask when certain methods might be most effective.

COMMAND LANGUAGE VERSUS DIRECT MANIPULATION

In the HCI literature, two types of interaction styles are generally recognized: command language or direct manipulation systems. Command language systems are also known as linguistic manipulation systems, or dialogue systems, and were often used in the early days of computers when users communicated with the computer through text command. Direct manipulation systems are the graphic user interfaces (GUI) now common to users in the Windows environment. Shneiderman is credited with introducing "direct manipulation" as a phrase for interfaces with the following characteristics: continuous representation, physical actions instead of typed commands, and the rapid impact on objects with the results becoming immediately visible (Helander, 1998).

Shneiderman (1997) argues that the usefulness of direct manipulation stems from the visibility of the objects of interest so that there is little need for the mental decomposition of tasks into multiple commands. Each action produces a result in the task domain that is visible in the interface. He relates the basic principle to stimulus-response compatibility discussions in the human factors literature. He claims that the difficulty with direct manipulation is to come up with an appropriate representation or model of reality (Shneiderman, 1987). We will see later how this discussion is renewed in the research literature on metaphor and simulation.

HYPERTEXT

Hypertext is an important issue in HCI research, and we will see throughout this book how central it is to discussions of designing educational software. It is connected to the literature on cognitive issues because hypertext is said to mimic the associative manner in which the brain works. In the past it has been argued often that hypertext may alter the way people read, write and organize information, and may be crucial in the development of non-linear thinking (McKnight, Dillon & Richardson, 1991). This literature claims that linear text limits an author's ability to address the range of needs and interests of readers. Hypertext solves this problem, the argument goes, by presenting text in a non-linear arrangement linked by key phrases in the text (Osgood, 1994). Additionally, in line with the media discussion in the second part of this book, one of the most important advantages of hypertext is that it is a method for integrating three technologies and industries that have been separate until recently: publishing, computing, and film and television broadcasting (Nielsen, 1990).

However, as discussed at length in Chapters Five and Eleven, the literature shows that one of the primary problems with hypertext is that it causes severe difficulty with navigation for users (McKnight, Dillon & Richardson, 1991; Osgood, 1994). Additionally, hypertext indexing methods are often inadequate and not necessarily focused on what the user most wants to follow. Researchers recognizing the problem of navigation, work on how to help users better navigate through text including more effective forms of indexing and the use of narrative schemes (discussed in Chapters Five, Nine, Ten and Eleven).

Some argue that the opposition between hypertext and print reading is a false dichotomy. McKnight, Dillon and Richardson (1991) point out that reading is not really a linear activity, but instead involves a great deal of skimming. Particularly in experienced readers, rarely is a document read straight through from beginning to end. Consider how one typically reads a newspaper, moving from one article to another, page to page in a non-linear fashion. The problem with hypertext is that its theoretical basis, which is an implied criticism of normal text forms, is inaccurate, and consequently the alternative is not the advantage proponents imagine. While hypertext represents a change in the presentation of text, it may not alter how words are read by a reader.

GRAPHIC AND VISUAL ISSUES

In addition to visual interface issues, the HCI literature also touches on topics related to visual perception and how the specifics of human visual perception may impact human-computer interaction. The important issues include how light transmits information to the eye of the perceiver, how that information is processed, and how that information results in conscious experience of the external world. The notion of the perceiver as a processor of information is the central focus of the psychology of visual perception (Haber & Hershenson, 1973). Some interested in broader visual research have examined the relationship between visual imagery and mental imagery in human perception (Klima, 1974).

Rudolf Arnheim's work (1974) is central (and often cited) in this discussion of perceptual issues in HCI. He argued that "gestalt," the German word for shape or form, has been applied since the beginning of the 20th century to a body of scientific principles that were derived mainly from experiments in sensory perception. Arnheim points to Christian von Ehrenfels' claim that the sum of the experience of twelve observers listening to one of the twelve tones of a melody is quite different from the experience of someone listening to the whole melody. He argues that in a similar manner, vision is not a mechanical recording of individual elements, but rather the recognition of patterns. Consequently, much of the research has focused on visual pattern recognition.

According to Shneiderman (1997), visual perception is underutilized by today's graphical user interfaces. His work on the HomeFinder and the FilmFinder demonstrated that users could find information faster with graphical user interfaces than with natural language queries, and that user comprehension and satisfaction was high for these interfaces. Furthermore, the literature suggests that there is evidence to support that humans recall pictures better than words (Faulkner, 1998).

This research on visual issues is important in educational environments, particularly Arnheim's recognition of meaning making through pattern recognition. In constructing educational software using visual images, designers should be aware of the tendency of viewers to draw meaning from whole patterns, rather than individual elements.

METAPHOR

Interface metaphors are often discussed in HCI literature as they pertain to interface design. The use of an interface metaphor—such as the desktop and window—is widespread in computer software design as an ideal method for providing a quick and easy foundation for users to understand how applications work (Cohen, 1997). Interface metaphors work by exploiting previous user knowledge of a mental model (Helander, 1998; Klima, 1974). According to Helander (1998), there are three main approaches to metaphor research: measuring behavioral effects, cognitive mappings between metaphor and meaning, and the constraints of context and goals when using particular metaphors.

In the literature on interface metaphors, critics claim that metaphors stand in the way of making new connections and associations (Nelson, 1995). Research in cognitive psychology supports this notion that using similar representations is helpful, but can be detrimental to user behavior under specific conditions, particularly if the metaphor does not fit appropriately (Cohen, 1997). Nelson feels that metaphors are counterproductive because they keep designers from finding new design principles that might lead to a new conceptual organization. In a similar fashion, Oren (1995) sees the use of metaphor as a genre where familiarity to images and conventions prevent users from taking a more active role.

The use of interface metaphors in educational software is utilized from general online campus icons, to course tools such as blackboards paralleling the traditional classroom. However, we see the research literature suggests that this may be limiting and work against learning. This is particularly useful to understand when working as educators to create new learning environments.

ANIMATION

Animation is another subject discussed in the research literature, usually addressed along with interface design issues. The term "animation" is not used to describe drawn figures, but rather to describe movements of either text or graphics on the computer screen. It is the use of graphic art occurring over time (Baecker & Small, 1995). Animation is not used as much as it could be in human-computer interactions. Many in the literature argue that it can be very effective in establishing mood, in increasing a sense of identification in the user, for persuasion, and for explication (Baecker & Small, 1995; Morris, Owen & Fraser, 1994). Baecker and Small (1995) describe many specific uses for animation including reviewing, identifying an application, emphasizing transitions to orient the user, providing choices in complex menus, demonstrating actions, providing clear explanations, giving feedback on computer status, showing history of navigation, and providing guidance when a user needs help.

In terms of assessment, there is disagreement about the effectiveness of animation. Morris, Owen and Fraser (1994) claim that several studies have explored the effectiveness of animations in educational contexts, while Bederson (1998) argues there have been few studies providing clear evidence of the positive affects of animation for the user.

ORGANIZATIONAL ISSUES

The literature on HCI also addresses issues having to do with how computers are used in organizations. Increasingly, HCI researchers are looking at not just the individual characteristics of the user, but at interactions among people mediated by computers (Faulkner, 1998; Malone, 1987; Wooffitt, Fraser, Gilbert & McGlashan, 1997). Rather than focusing on the user, this approach looks at groups of users, and at how to design computer systems in such a way that they fit naturally and appropriately into human organizations.

Malone (1987) identifies four basic aspects of the organizational issues in HCI: economic, structural, human relations, and political. Maddix (1990) sees the emphasis on organizational HCI analysis as rising as organizational changes lead to workgroups characterized by a collective mission instead of individuals. In fact, some argue that differences in users' interactions with systems are not the result of individual psychological and physical differences but social structure differences (Wooffitt, Fraser, Gilbert & McGlashan, 1997).

As the research literature supports group learning, the dynamics of group interaction is vital to understand. Software that accommodates group behavior and typical social structures is very important for educational software.

TASK ANALYSIS

Jonassen, Tessmer, and Hannum (1999) argue that instructional and assessment strategies should vary with the nature of the learning outcome. Consequently, task analysis for instructional design is a process of analyzing and articulating the kind of learning that one expects the learners to know how to perform.

Although task analysis emerged as a process in the behaviorist era of instructional design, task analysis methods have followed the paradigm shifts to cognitive psychology and on to constructivism. According to Jonassen, Tessmer and Hannum, task analysis consists of six distinct functions: classifying tasks as learning outcomes, inventorying tasks, selecting tasks, decomposing tasks, sequencing tasks and task components, and classifying learning outcomes.

For educational applications one needs to consider the degree to which learning can be broken down into tasks. Not all subjects fit this model. Additionally, how does tasks analysis connect with assessment of student learning? Some tasks may not lend themselves to assessment.

CONCLUSION

In this chapter, we reviewed the research literature on human-computer interaction as it impacts educational applications. The related fields of human factors and usability were examined, as well as the key interface design research. The GOMS model and the distinction between command language and direct manipulation were reviewed. The more expansive issues of hypertext and metaphor use were raised and discussed in relationship to educational applications.

Now that readers have a broad understanding of the general human-computer interaction field, we can examine in depth specific applications of computer use in educational environments. While interactivity is certainly an important issue in HCI literature, it is central to a discussion of creating learning environments. Navigation through computer environments impacts learning in surprisingly complex ways. The next chapter looks at these issues in detail.

Chapter VI

Interactivity and Navigation

Interactivity is a key issue in the designing of educational software. Critics of computer-based education often claim that these courses lack the necessary face-to-face interaction required for learning. In this chapter we address two distinct types of interactivity: student-to-student and teacher- and student-to-media. In practice, computer-based courses vary greatly in the amount of interactivity incorporated in the programs. In terms of student-to-media interactivity, some computer-based training is simply a slide show with interactivity only available to the students through their role of clicking on an arrow to move to the next slide. Other software programs have very large amounts of interaction, giving users control through simulations, communication with other students and the instructor, and even structuring the environment through simple computer programming. In fact, anecdotally, distance learning faculty and students consistently report that there is more interaction in this latter group than in face-to-face courses. In relation to human interactivity (student-to-student and student-to-teacher), some computer-based courses incorporate no interaction and only use faculty to evaluate and grade student work, while other distance learning courses have extensive human interaction. For software designers, there are very important questions about interactivity. First, how much interactivity should be used or encouraged? Second, how is interactivity different using various media? Third, what are the differences and similarities of interactivity in computer-based educational environments as

Interactivity and Navigation 71

opposed to the traditional classroom? Fourth, what does media theory tell us about interactivity and effects on users that is particularly relevant to education?

The 2001 survey showed that students felt the quality of interaction in distance learning courses between the instructor and students was comparable to that found in face-to-face courses, with 74.6% of the respondents from computer-based courses strongly agreeing and agreeing with the statement, "The quality of the interaction with the instructor was the same or better as in traditional face-to-face courses" (Figure 12).

Note that the videotape format courses fared much worse in this student evaluation, with only 33.8% of the respondents strongly agreeing or agreeing with the statement.

In terms of student-to-student interaction, the 2001 survey found that there was slightly higher disagreement (55.9% for computer-based courses; 82.5% for videotape) with the statement, "The quality of the interaction with other students in the course was the same or better as in a traditional face-to-face course" (Figure 13).

Figure 12: Quality of Interaction with Instructor by Delivery Format (Questions 1 & 3)

Delivery format * The quality of the interaction with the instructor was the same or better as in a traditional face-to-face course. Crosstabulation

			The quality of the interaction with the instructor was the same or better as in a traditional face-to-face course.				Total
			strongly agree	agree	disagree	strongly disagree	
Delivery format	computer-based	Count	21	23	12	3	59
		% within Delivery format	35.6%	39.0%	20.3%	5.1%	100.0%
	videotape	Count	3	18	27	14	62
		% within Delivery format	4.8%	29.0%	43.5%	22.6%	100.0%
	correspondence	Count	2		2	1	5
		% within Delivery format	40.0%		40.0%	20.0%	100.0%
	other	Count	1				1
		% within Delivery format	100.0%				100.0%
Total		Count	27	41	41	18	127
		% within Delivery format	21.3%	32.3%	32.3%	14.2%	100.0%

Figure 13: Quality of Interaction with Students by Delivery Format (Questions 1 & 4)

Delivery format * The quality of the interaction with other students in the course was the same or better as in a traditional face-to-face course. Crosstabulation

			The quality of the interaction with other students in the course was the same or better as in a traditional face-to-face course.				Total
			strongly agree	agree	disagree	strongly disagree	
Delivery format	computer-based	Count	6	20	22	11	59
		% within Delivery format	10.2%	33.9%	37.3%	18.6%	100.0%
	videotape	Count	1	9	27	20	57
		% within Delivery format	1.8%	15.8%	47.4%	35.1%	100.0%
	correspondence	Count	1	1	2	1	5
		% within Delivery format	20.0%	20.0%	40.0%	20.0%	100.0%
	other	Count		1			1
		% within Delivery format		100.0%			100.0%
Total		Count	8	31	51	32	122
		% within Delivery format	6.6%	25.4%	41.8%	26.2%	100.0%

Clearly the respondents feel that the quality of the interaction between student and instructor is better than the interaction among students. This is an important finding and points out the problem particularly in courses that use a tutorial method and miss this aspect of interactivity found in the classroom. Nevertheless, administrators seem to have a different viewpoint than students on this matter. According to the 2000 survey, most administrators (44.7% "strongly agree," 41.2% "agree") claim that their courses have a significant amount of interaction among students (Figure 14).

I surmise from this that while administrators understand the importance of student-to-student interaction, their attempts to provide significant interaction may not lead to providing the same level as face-to-face courses.

Many of the administrators confirmed this focus on human interaction issues.

> Well, that's always been part of our approach to sit down with the faculty member and figure out how to do that [incorporate interactivity]. Our courses have a great deal of faculty-to-student and student-to-student interaction. For example, in the course I teach we tried to enroll a large number of students, but when you have 100 or 150 students the amount of interaction with the students is minimal. So we've set up guidelines so that, for example, in my course for every thirty students another faculty member is assigned (Vice-President, anonymous large, independent, Eastern U.S. doctoral degree-granting institution).

One administrator commented on the problem of instructors not interacting to a sufficient degree with students.

> In terms of the online, that's one of the things we try to watch, how much interactivity is there. I have an issue right now, a student filed a lengthy complaint about an instructor who was not being responsive, not being

Figure 14: Significant Interaction (Question 41)

Courses include significant interaction with other students.

		Frequency	Percent	Valid Percent	Cumulative Percent
Valid	strongly agree	76	43.2	44.7	44.7
	agree	70	39.8	41.2	85.9
	disagree	20	11.4	11.8	97.6
	strongly disagree	4	2.3	2.4	100.0
	Total	170	96.6	100.0	
Missing	System	6	3.4		
Total		176	100.0		

interactive, and we jump on those kinds of complaints (Thornton Perry, Director of Distance Education, Bellevue Community College).

One administrator spoke about how she sees her job as focused on facilitating interactivity.

My job is to help students and faculty become better acclimated to a technology base, with issues such as interaction, and learning styles, that they swear they know, but it isn't until they get into a videoconference classroom that they really understand (Jacquelynn Sharpe, Division of Distance and Distributed Learning, Georgia State University).

Some respondents claimed that there is more student-faculty interaction in distance learning courses than in face-to-face ones.

Let me take student-faculty interaction first. We are finding that students say in these courses that they are interacting much, much more with their instructor than in face-to-face. And if you think about it, it makes sense— you require a lot of feedback, everyone has to be in on this discussion list twice a week, here's the three questions. So there's a lot more interaction. I always say my first distance learning course was a sociology course I took with 500 people in a lecture hall. I never interacted with that instructor. With our online courses we are getting quite a bit of interaction (Greg Chamberlain, Dean of Learning Resources, Bakersfield College).

Most of our students say that they have more interaction with each other and the faculty mentor than in classes on campus. All of our faculty mentors have 800 numbers, so regardless whether a student lives in Kansas, Missouri, Oklahoma, or wherever, they can always get in contact with their faculty mentor by 800 number. So, it is at no cost to the students. Also, all the faculty have home computers with access to email. We also require students to have computers with access to email. So there's a great deal of interaction between faculty and students (Joy Edwards, Director of Graduate Studies, Texas Wesleyan University).

With telecourses, I've had very few complaints over the years about interactivity. In the case of my course, they have forty hours of television to watch. And they can still get a hold of me by phone or email. They can get more instructional contact hours potentially, than if they were sitting in a class (Thornton Perry, Director of Distance Education, Bellevue Community College).

Many noted the differences in interaction among delivery platforms and technologies used. Notice in the following interview how an administrator described in-person meetings used to increase interactivity.

> The video courses I consider to be interactive. We encourage faculty, and we pay them, to travel to the remote sites at least once or twice. We have the campus facility and four remote sites, usually a faculty member will go to those sites at least once or twice a semester. We'll pay them to do that. For the correspondence courses, I'd say 97% of the students live close by....They can contact the faculty member via phone, email, or office hours. Unlike traditional correspondence courses, we don't have any students living out of state. Most of the students live within sixty miles of the campus, we encourage them to come on campus, most come here to test. So that's how we encourage interaction (John Burgeson, Dean, Center for Continuing Studies, St. Cloud State University).

One respondent spoke about specific techniques used for interaction driven by faculty decisions.

> There are only two instructors who deliver their courses via email. Currently, eight-five out of ninety of our online classes are delivered via Etudes, a course management system developed by one of our faculty. Instructors use the Dialogue Chamber (within Etudes) for posting questions and class discussions. Students can talk amongst themselves, publicly in the Dialogue Chamber or privately via the Private Message Center, an email system built into Etudes. The instructor and students can communicate privately via the private message center as well. There is also a chat room for synchronous discussion. Very few instructors use the chat feature, however (Vivian Sinou, Dean, Distance & Mediated Learning, Foothill College).

In the next interview the administrator emphasizes the student-to-student interaction at a single corporate site where a group of employers are taking a course together.

> You are really asking a question that reflects where industry is going. Different delivery methods promote different modes of interaction. In the

sense that if you are taking a microwave course you may be the only one at that site, or there may be a number of students at that site. The course is broadcast live so the students come together at the same time, watch the broadcast, they can discuss it, and if the company wants it they can get the capability to ask questions during the class. That's as close as you get to being in the class, and you have your colleagues there, potentially, depending on how many sign up. You can have the same sort of thing, except that it is not at the same time that the class is occurring (Program Manager, anonymous large, independent, Western U.S. doctoral degree-granting institution).

Providing sufficient interaction seems to be the biggest problem for telecourses and two-way video courses.

The area that I think we have the least student interaction is with our telecourses. The interaction is just a few review sessions....But it isn't geared for that. It is geared to the person who wants all the information, now I can take the test and be done with it (Greg Chamberlain, Dean of Learning Resources, Bakersfield College).

One administrator described a strategy to address this problem by incorporating face-to-face time with the instructor.

With the two-way video the interaction tends to be nil outside of class. We've tried to mitigate that by saying go to the remote site and teach backwards. Gives them a chance to meet the students, and to know that they do exist below the legs, that they do have legs (Greg Chamberlain, Dean of Learning Resources, Bakersfield College).

Increasingly, a blended approach using some face-to-face is common in higher education. A prime example is the new University of Phoenix FlexNet program where students meet face-to-face for the first and last meetings, and complete the rest of the course (three meetings) online. In some instances, this blended approach is brought forth by accreditation issues and may simply be an evolutionary stage. In other cases, the initial and ending contact may prove to have an important pedagogical impact, particularly for certain audiences and subject matter where learning is improved by face-to-face interaction. Some administrators discussed this strategy where some in-person contact is required.

> There is required face-to-face for the telecourses. They are required to come to campus to take the test, they come for orientation, they come for at least one review session. The two-way video, obviously that's face-to-face (Greg Chamberlain, Dean of Learning Resources, Bakersfield College).

> The faculty requires at least two campus visits for each course—to take a midterm exam and for a final examination. Many faculty also offer review sessions, so there is a lot of face-to-face contact provided that way. We are offering science classes with lab work on Saturday mornings. It depends on the course (Arthur Friedman, Professor and Coordinator, College of the Air, Nassau Community College).

> It varies. The online and correspondence type courses can be completely without face-to-face time. The video is the same way, a one-way method. You do have the email and voice mail as well. Some of the videoconferencing require campus visits. For the most part our emphasis is not on that. We've tried to make our program as true to distance learning as possible (Allan Guenther, Marketing Coordinator, Distance Education, The University of Alabama).

Nevertheless, some administrators reported the need for completely distant courses to meet the market and student needs.

> No, we haven't used that approach because our students are so geographically dispersed that it wouldn't have been practical. We do have programs which require them to come on campus at the beginning, but that essentially excludes the international student. The only time we encourage them to come on campus is for graduation. But I do think that option is important, but it should not be required (Vice-President, anonymous large, independent, Eastern U.S. doctoral degree-granting institution).

> The online courses to this point have been geographically boundary-free. They do not need to come to campus. In fact, we've finally got our matriculation process down to the point where they can go through the whole matriculation without coming to campus (Greg Chamberlain, Dean of Learning Resources, Bakersfield College).

As with many aspects of distance learning, one respondent stated that interaction is a faculty responsibility.

That's at the discretion of the faculty member (Program Manager, anonymous large, independent, Western U.S. doctoral degree-granting institution).

Some of the administrators put the responsibility on the students to interact in the manner, and to the degree, that suits them.

Up until the mid-nineties, when email was used, before that it was the old-fashioned phone calls and the opportunity to meet in person in the classroom, or to come to office hours. But it has been the student who needs the convenience because of family, job, travel restrictions (Thornton Perry, Director of Distance Education, Bellevue Community College).

That has been a problem in our video and audio courses. I'll tell you what we've done there, and then where we're going with the new online courses. Typically on the audio and video courses, the faculty are no longer with us, so we have a grad[uate] fellow do the grading. So the student is encouraged to contact the grader by email, mail, and phone. The grader stays in contact with the student to keep on top of grades and that sort of thing so that there is accountability on our end. However, to a large part it is up to the individual student's initiative to contact the grader. For the new courses, we are delivering content on CD-ROMs because of expense and iffy Internet connections internationally. So we will be giving them CD-ROMs with the ability to also communicate with graders through the Internet or by different means, and with the capability to also work in groups if there are a group of students who want to go through a course together (Jon Raibley, Assistant Director of the Center for Lifelong Learning, Western Seminary).

Oddly enough, not all distance learning students feel that interactivity is important. According to the results of the 2001 survey, more than half (52.4%) of the respondents did not agree with the statement, "I would prefer active interaction with the course material, instructor, and other students over recorded lectures and prepared materials."

While many assume that interactivity is a positive and necessary attribute of effective distance learning courses, clearly many students are drawn to a more solitary and passive learning experience. In Chapter Three, we saw a parallel response to the issue of students working in groups. Also, in Chapter One, Patricia Cross claimed that adults often do not want self-directed learning. One of the major

Figure 15: Prefer Active Interaction (Question 1 & 8)

Delivery format * I would prefer active interaction with the course material, instructor, and other students over recorded lectures and prepared materials. Crosstabulation

			I would prefer active interaction with the course material, instructor, and other students over recorded lectures and prepared materials.				Total
			strongly agree	agree	disagree	strongly disagree	
Delivery format	computer-based	Count	8	13	27	8	56
		% within Delivery format	14.3%	23.2%	48.2%	14.3%	100.0%
	videotape	Count	20	15	22	5	62
		% within Delivery format	32.3%	24.2%	35.5%	8.1%	100.0%
	correspondence	Count	1	2	2		5
		% within Delivery format	20.0%	40.0%	40.0%		100.0%
	other	Count			1		1
		% within Delivery format			100.0%		100.0%
Total		Count	29	30	52	13	124
		% within Delivery format	23.4%	24.2%	41.9%	10.5%	100.0%

issues that needs to be addressed is the conflict evident here between what might make best pedagogical sense in distance learning, and student learning preferences.

INTERACTION WITH MEDIA

The second type of interaction I want to focus on in this chapter is that between student users and the media or "hypermedia." Hypermedia is a term that describes the combination of media that are related in computer environments through programmed links. According to Lennon (1997), the term hypermedia was first coined by Theodor Nelson in a 1965 paper where he described it as films, sound recordings and video recordings arranged as non-linear systems. The advantage of computer hypermedia compared to traditional educational media is that it is non-sequential and able to present material in a non-linear fashion that could not be conveniently done on paper. Other traditional technologies such as blackboards cannot easily store information. Lennon argues that many new ways of using old media as hypermedia does yield entirely new techniques and communication facilities. What are the benefits of using old media in hypermedia contexts?

Chan and Ahern (1999) argue that "flow theory," a psychological theory of optimal performance (Csikszentmihalyi, 1991), can be useful in instructional design because it provides several structural variables that can be manipulated in order to increase the likelihood that a learner will be motivated to learn. In their study, they found that extensive use of multimedia can work against flow and a learner's motivation to continue. They suggest that the use of multimedia increase in applications as a user becomes less challenged and more familiar with the course material.

In educational environments, it is important to understand how students interact using text or hypertext. Michael Joyce (2000) defines reading on the computer as the difference between "looking at and looking through." His literary critic point of view focuses on how readers are turned into writers by navigating through hypertext space. Joyce argues that hypertext requires users to become actors rather than passive viewers. He theorizes that the hypertext effect depends on a re-reading approach. In fact, reading hypertext recreates the author's experience of re-reading his or her own work. Joyce describes this repositioning as "othermindedness." According to him, hypertext is better suited to education, while hypermedia is more like television. A result of the use of hypertext in education may be that "successive attendings" are more useful than a long attention span to perceive multiple perspectives, variety, and access to knowledge.

One of the great challenges in designing software for student use is in finding effective ways for users to make their way through often confusing and disorienting environments. The 2001 study found that students overwhelmingly reported wanting control over course navigation (Figure 16). Respondents reported strong agreement or agreement with the statement, "I like maximum control over how to navigate through the course software" 91.9% of the time

In the 2000 study, 84.7% university administrators agreed or strongly agreed with the statement that great care is taken in understanding how students navigate through the computer software.

In interviews with the distance learning administrators, this concern for effective computer navigation strategies was evident. For those involved in the administration of distance learning courses, it is common knowledge that the problems with using software begin with simply logging into password protected learning spaces.

Figure 16: Student Control of Navigation (Questions 1 & 20)

Delivery format * I like maximum control over how to navigate through the course software. Crosstabulation

			I like maximum control over how to navigate through the course software.			Total
			strongly agree	agree	disagree	
Delivery format	computer-based	Count	19	25	4	48
		% within Delivery format	39.6%	52.1%	8.3%	100.0%
	videotape	Count		1		1
		% within Delivery format		100.0%		100.0%
Total		Count	19	26	4	49
		% within Delivery format	38.8%	53.1%	8.2%	100.0%

Figure 17: Attention to Navigation Issues (Question 44)

Great care is taken in understanding how students navigate through the course software.

		Frequency	Percent	Valid Percent	Cumulative Percent
Valid	strongly agree	66	37.5	39.1	39.1
	agree	77	43.8	45.6	84.6
	disagree	19	10.8	11.2	95.9
	strongly disagree	7	4.0	4.1	100.0
	Total	169	96.0	100.0	
Missing	System	7	4.0		
Total		176	100.0		

The only technical questions we get relate to login difficulties, such as students forgetting their passwords, inability to figure out their user IDs, and the like. A common question is, "I don't know my password. How do I get it?" (Vivian Sinou, Dean, Distance & Mediated Learning, Foothill College)

There have been problems with getting passwords for students who never come to campus. They try to use their password, and every time they are bounced out. The university has come a long way in terms of student access and support in the past three years. All of these functions are available online for students whether they are residential or remote. That's where the Student Support Services Group comes in because they have to troubleshoot, find out what is wrong, and get them into the course. That has been more challenging than the whole layout, the look and feel of the courseware itself (Carole Hayes, Coordinator, External Relations and Development, Office for Distributed and Distance Learning, Florida State University).

Although Blackboard is pretty intuitive, students are having problems with passwords and other mechanical tasks. So I've developed a front end on a separate web page to walk them through that. There will be a tutorial, browser checker, a quick start page where they click on one button and it takes them to the class directly. There is another button that if they click it and type in their email address it will come to me and tell me that they are having problems and need help. As a Blackboard administrator, I can change their password back to the default, so within a few hours they can be back online. So I try to take away that problem for them (Don Cardinal, Chapman University, School of Education Faculty member).

Additionally, the same respondent spoke about a technique to prepare students for taking online courses.

> I've also written an email to the faculty about online courses suggesting that students need to begin the courses before the start date. Before they go through day one, they need to visit the course web site and go through a tutorial, make sure it is set up properly on their computer, that the hardware needs are set (Don Cardinal, Chapman University, School of Education Faculty member).

Most of those surveyed responded that they use off-the-shelf software such as Blackboard and WebCT. Although they sometimes customize the software to some degree based on the course content, the general navigation scheme is set by these licensed platforms. One administrator described enhancing licensed software with proprietary code.

> Yes, also enhanced by a locally grown software called WebMC. It's one that the faculty like very much, it's a very easy authoring tool. It coordinates well with Blackboard. So faculty members use this software which has been around for some years. They can link within Blackboard to any work they have done in WebMC (Carole Hayes, Coordinator, External Relations and Development, Office for Distributed and Distance Learning, Florida State University).

Another administrator described how courses are standardized through the use of templates and individualized to some degree by context.

> We have a basic template for each course. But if you look at each of our courses online you'll see they have a very different look online depending on their course content. They all have login and certain headers, but then on the flip side they all have their own look and feel based on the course content (Allan Guenther, Marketing Coordinator, Distance Education, The University of Alabama).

Some institutional representatives describe their own proprietary platforms. But in using this courseware which is pretty flexible, the student

> pops into the front page where there are announcements, and there are buttons all along the side which have syllabus, faculty and staff information, grading policies. On the front page, they have everything. In terms of usability, most people are pretty happy with it (Carole Hayes, Coordinator, External Relations and Development, Office for Distributed and Distance Learning, Florida State University).

> The layout of Etudes is straightforward and easy to navigate. It consists of a classroom, assignment desk, forums, chat, and a testing center. Students don't have to go to their email for their coursework. They can do everything within the platform. The assignment desk is where students turn in their assignments. They can take their tests in the testing center. It's similar to other course management systems. It's very easy to use (Vivian Sinou, Dean, Distance & Mediated Learning, Foothill College).

Note here that the features these two administrators describe are very similar to the standard licensed software.

Many administrators emphasize faculty control of the course design.

> The professors have a number of options. They can set up their own website at no cost, but that requires that they know html, or their teaching assistant knows it, or I can make it for them. If I do, it's not going to be a very fancy one. They also have Blackboard and WebCT available to them (Elizabeth Spencer-Dawes, Manager, Distance Learning, Boston University).

> Any faculty member who wants to use WebCT for his/her class can, this is not just for the College of the Air (Arthur Friedman, Professor and Coordinator, College of the Air, Nassau Community College).

> If you asked ten faculty, you'd get ten different answers (Carole Hayes, Coordinator, External Relations and Development, Office for Distributed and Distance Learning, Florida State University).

Some comments reveal how haphazard the process of working with faculty appears to be.

> Most of them come to me at the last minute, so I just quickly put something up with links to pdf [portable document files] files so that they can

download documents (Elizabeth Spencer-Dawes, Manager, Distance Learning, Boston University).

We do have one faculty member who is the faculty coordinator, who goes all out with a professional web site, with streaming video and all, but he's unusual (Elizabeth Spencer-Dawes, Manager, Distance Learning, Boston University).

As few institutions have the resources to address the navigation issue in a concerted way, they end up relying greatly on the software vendors. This is a serious limiting factor at present. Nevertheless, increasingly these companies will need to address pedagogical issues. Lotus Learning Space is one piece of software that has extensively thought about building in collaborative opportunities with software, particularly for use in corporate training environments. Unfortunately, this platform has not been widely adopted in higher education at this point because of cost and the requirement for students to purchase software. As a result, at this time the vendor seems to have cut back attempts to market to educational providers.

The literature on the subject of navigation in education environments points to many practical issues. Sumner and Taylor (2000) argue that students should always be able to answer questions during the courses such as what they need to do, why they are doing it, and when is it due. Furthermore, they point out that distance learning environments need to support orientation, understanding the relationship between tasks and resources, establishing and maintaining new study habits and ways of working, confidence building, integration, enrichable and annotateable learning materials, tracking and feedback. Bourdeau (1999) argues that there is a problem in making transitions between individual, team, and group media spaces that interferes with learning experiences by causing disorientation, loss of time and attention.

Many in the research literature have commented on the problem of users getting lost in navigating their way through hypertext environments. David Jonassen (2000) points out that the greatest problem with hypermedia is integrating the rich information into learner knowledge structures. For this reason, it is more effective for students to create hypermedia than to use such documents. When creating hypermedia documents, students are forced to make sense of the knowledge they are gaining. However, Jonassen notes that the biggest problem with this approach is that it is time-consuming and expensive.

Lydia Plowman (1996) claims that navigational practices in computer programs work against the coherence of the learning experience, causing it to be very fragmented and disorganized. For Plowman, user control of the narrative experience is reassuring, but confusing. The demands of interactivity requiring interaction in order to progress, the simultaneous use of multiple media, and the stress of interpreting icons and other interface conventions are all examples of practices in computer-based educational programs that interfere with learning. Plowman claims that one of the main problems is that interactivity at this point in the development of technology is "meager" in comparison to everyday human interaction. Real classroom interaction involves opportunities for customization and the addressing of specific learning problems on an on-going basis. In addition, interactivity sometimes works against learners in that they are less likely to reflect on tasks before acting.

Johnson (1997) traces the rise of the 19th century novel to the function of serving to explicate new social realities, while the 20th century computer interface addresses the overwhelming amount of information in search of meaning. Thus narrative in computer environments becomes a mix of metaphor and footnote and is atomized. Johnson concludes that interface design should be organized around meaning instead of space. What are the educational implications of organization around meaning?

Petre, Carswell, Price and Thomas (2000) claim that in distance learning, course delivery it isn't enough to simply translate existing practice; the underlying function must be served through strategies that exploit the strengths of the medium. What are these strengths?

CONCLUSION

In this chapter, we looked at how students and educators view the issue of interactivity in computer environments. The research literature reveals that organizing and understanding the effects of interactivity on students is a primary pedagogical concern. According to Jonassen (2000), social navigation of the Internet occurs when information is shared among users to assist in making sense of the computer space. How can this social navigation be encouraged and managed in educational environments? One approach commonly used is to provide threaded discussion areas for students in online environments to share navigation information. What other approaches might be helpful?

What does film theory say that might apply to more effective navigation approaches? Can transition principles from film be used in computer environments? Since viewers are accustomed to the coding of film and television, might not these principles be useful for controlling navigation? Specifically, how might this work? We will look at this issue later in second part of the book when we turn to the computer as medium.

In this chapter, the issues surround the important questions of interactivity and how users navigate through computer learning environments were addressed. In the next chapter, I look at specific computer tools to increase student learning.

Chapter VII

Computer Tools for Learning: Concept Maps, Mindtools, Computer Programming

One way of looking at the advantage of the use of computers in education is to see them as assisting with the organization and development of thinking structures in the mind of students. Concept maps offer one specific answer to the challenge of creating effective navigational schemes for educational software, while at the same time facilitating the organization of the user's thinking. Mindtools are another set of software tools that assists learners in their thought processes. Additionally, some have argued that the process of learning a computer language in and of itself leads to important cognitive development in the student. In this chapter, I consider these specific tools and the educational purposes they serve.

CRITICAL THINKING

When one talks about these various approaches to the use of computers in educational environments often the discussion centers on a supposed advantage in teaching students higher order types of thinking, sometimes called "critical thinking." However, the concept of critical thinking lacks a common definition and most higher education institutions have not defined or assessed critical thinking in a substantive way. Cook et al (1975) define critical thinking as the "reflective, systematic, rational, and skeptical use of cognitive processes." Others see three major components of critical thinking: knowledge, thinking skills, and attitudes (Aretz, Bolen, & Devereux, 1997). However, when looking at specific assessment efforts, critical thinking is a complicated notion and probably cannot be measured in a single way.

A great deal of inference is involved in the measurement of higher order thinking skills generally because they are not easily quantifiable (Reeves, Laffey, & Marlino, 1996). Furthermore, cognitive assessment is difficult to perform because it is relatively easy for students to mimic conceptual knowledge without really understanding the web of meaning. Rather than trying to make large claims about a learner's overall critical thinking or cognitive abilities, concept maps, mindtools, and computer programming might assess a learner's ability to make meaningful connections between individual ideas and concepts, and both encourage the development of higher order thinking and reveal patterns of student thinking or cognitive structure.

CONCEPT MAPPING

A concept map is a graphical representation of the cognitive structure of an individual learner—a map of how the learner sees the different elements of a given subject and their interrelationships. Concept maps have potentially great utility in the distance learning delivery format because they are graphic representations and are built on a cognitive model in line with the associative thinking used in hypertext. The idea behind hypertext is that the human brain operates in an associative manner in moving from one thought to another. These associations and their network of connections can easily be drawn in a concept map form. Furthermore, concept maps are used both in instructional delivery and in assessment, and thereby may serve as an integrated model for distance learning delivery in the future. In fact, there are already efforts to create graphic user interfaces (GUIs), based on concept maps. Since academic quality is a continual issue in distance learning format courses, measuring student learning outcomes is very important. Furthermore, as distance learning continues to expand, the importance of analyzing higher level thinking outcomes is going to increase.

Concept maps, also known as semantic networks and mind maps, are spatial representations of concepts and their interrelations as stored in the human mind. The cognitive theory underlying concept mapping is known as semantic networking theory which hypothesizes that human memory is organized according to meaningful relationships between ideas in memory (Jonassen, Reeves, Hong, Harvey, &

Figure 18: Concept Map

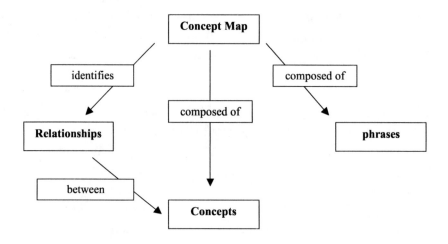

Peters, 1997). Students often lack an understanding of connections among concepts and therefore have difficulty in applying knowledge. Furthermore, knowledge must be organized in order to be retained. Students learn through assimilating new propositions into their cognitive networks, altering existing propositions, and altering the network as a whole (Schau & Mattern, 1997). Mapping techniques using graphical representations of interconnections are useful in this regard as both a learning tool which can increase long-term memory and as an assessment instrument. According to Jonassen (2000), concept maps help to organize learner's knowledge by integrating information into a progressively more complex conceptual framework. They help to increase the total quantity of formal content knowledge because they facilitate a learner's ability to use the skill of relating patterns and relationships. However, Jonassen recommends that students, not teachers, construct concept maps to encourage student concentration on process. Some have looked at concept mapping as a way of getting learners to focus on their own learning process, or as a metacognitive strategy. Metacognitive strategies are those that empower the learner to take charge of her/his own learning in a highly motivated fashion (Novak, 1998). In this manner, concept maps are very useful for self-assessment. Thus concept mapping has the advantage of being both a learning and assessment tool.

Concept maps are used extensively for assessing learning outcomes. As assessment instruments, concept maps can reveal a learner's cognitive structure. Assessing a learner's cognitive structure is an important indicator of learning because research has shown that during the process of learning, the learner's knowledge structure changes and becomes similar to that of the instructor (Jonassen et al., 1997). Consequently, concept maps are good indicators of performance. Furthermore, assessment of connected understanding is an important educational goal and needs to be assessed (Schau, Mattern & Weber, 1997). The use of concept maps in assessing knowledge of conceptual interrelation normally includes two steps: elicitation and representation. Structural knowledge may be elicited from learners using free association, while elicitation methods normally focus on describing the interrelationships.

Nevertheless, as connected understanding is a complex activity, it is difficult to assess accurately with concept maps alone. Concept maps provide clues for appropriate instruction and can be highly personalized. This customization in fact represents one of the problems in using such assessment instruments—there can be multiple "right" answers to such mapping. It is also difficult to arrive at a universal scoring method for concept maps. This is partly because the maps allow the user to individualize his or her learning experience, and this makes comparison among students difficult. Typical uses of concept-mapping assessment are as pre- and post-tests. Some have experimented with giving various amounts of information to the learner in the map, either through filling in boxes or in giving a list of terms to enter. Scoring of concept maps is often done by comparing learner maps with the instructor-generated map. Obviously, this method assumes that there is only one ideal map. Furthermore, structural knowledge changes depending on time, context, and purpose and therefore has limited reliability as a one-time assessment instrument. Networks in the mind are very complex and therefore should include thousands of concepts recorded over a long period of time. Usually this is not done with concept map assessment. Jonassen et al. (1997) point out that concepts are linked to one another in multiple ways beyond a two-dimensional representation, so that the maps are inherently a simplification.

Concept mapping tools can provide a quick assessment of a learner's cognitive structure in relationship to a particular subject. However, there are two concrete limitations to the use of concept mapping: First, because it is best used as a qualitative measurement and requires instructor time and expertise to analyze, it may be impractical in many situations. Second, some students require time to understand the idea of how to take the concept-mapping test. Finally, the ability to use this higher order assessment instrument in a computer environment gives it great potential value. One advantage of concept mapping in a computer environment is that it might work as a self-assessment, or even as an interface for online courses. Computer-networked environments are beneficial in assessment because they can automatically record process behaviors of the user/learner. Nevertheless, to work well in a computer environment, this assessment tool needs to reduce faculty time, be more easily understood by the student, and be used consistently over a period of time.

TEACHING WITH PROGRAMMING

One of the most high profile figures in the computer-enabled learning field over the years has been Seymour Papert from MIT, who has been a leader in the promoting the learning of programming languages. Seymour Papert's computer language, "Logo," is designed to give children an opportunity to create simple computer programs by moving a digital turtle in different directions on the computer screen. He emphasizes a constructivist approach and puts forward the notion of "mathetics," by which he means the art of learning. His "constructionist" notion is that knowledge is design and argues for actively engaging students, rather than interpreting and encoding information. Papert believes that the use of simple programming languages allows students to learn by a process he terms "bricolage." Bricolage is a style of organizing thinking that is negotiational rather than planned in advance. By placing themselves inside the symbolic universe of computer programming and trying to move about, the students are put in close proximal relationship to their problem. To some extent, Papert probably overstates the value of learning Logo. However, the basic principle of using computer programming as a tool to visualize the thinking process is an important one.

The creation of expert systems can also be a useful kind of learning through programming. According to Jonassen (2000), expert systems are artificial intelligence programs designed to simulate expert reasoning to help decision-making. They are used in education to provide advice, tutoring systems, personal feedback, cognitive processing, and engaging critical thinking. Jonassen suggests that creating expert systems can be a good learning experience because it requires learners to break processes into logical parts. He identifies the problem with students learning by programming in the tendency to focus on learning one principle deeply, rather than learning a subject in a broad way. Further, Jonassen claims that the Logo software is difficult to learn, and the environment represents a very limited and simple problem. In contrast to Logo and other programming approaches, he claims that "microworlds" are exploratory learning environments with simulations of real worlds where students navigate and manipulate objects. Jonassen argues that effective microworlds are simple, but personally meaningful.

DiSessa (2000) argues that computers can be the technical foundation of an enhanced literacy. For her, understanding literacy involves understanding thinking, how material resources assist thinking, and what a literate culture means. Literacy involves external, materially based symbols and representations organized into subsystems with rules and conventions—much like computer programming. DiSessa claims that computer literacy means intelligence achieved cooperatively with external materials. Literacy is a social deployment of skills in a context of material support. Interestingly, she sees each specialized form of literacy, such as computer programming, as a genre. In fact, she defines literacy as the convergence of a large number of genres and social niches. We will see in the second part of this book how the use of genre in media might impact educational uses of the computer.

DiSessa claims that programming can lead to deep learning by creating "enrichment frames" to reconceptualize. For her, the specific characteristics of the medium are of great importance. Thinking comes with a medium, it is not simply a valueless vehicle for expression. DiSessa argues that the history of the teaching of programming shows that it was co-opted by vocational schools, and is consequently thought of as a vocation instead of a educational process or literacy. It is time to reconsider how teaching programming might be useful as a teaching approach, particularly in computer learning environments.

MINDTOOLS—COGNITIVE AMPLIFICATION

David Jonassen (2000) has made an important contribution to the use of computers in education by focusing on practical ways that widely available software can be used for learning. He makes the distinction between looking with and learning from computers, claiming that students learn from thinking in meaningful ways. Thinking engaged by activities through computers and by representing what students know in forms supplied by computers are two of the most important ways computers can be used in education. Jonassen points out that drill-type computer programs replicate old rote learning and that canned tutorial programs fail to address the complexity of human responses. Instead he argues for something he calls "mindtools," or computer-based tools and learning environments, that function as intellectual partners with the learner. They are "cognitive amplification and reorganization tools."

Unlike other theorists who concentrate on more exotic futuristic software, Jonassen writes about using current, widely available software such as databases and spreadsheets for educational purposes. He claims that database management tools help learners integrate and interrelate content, and that building databases requires learners to organize information by identifying the essential aspects of the content. Spreadsheets are an example of software that amplifies and reorganizes mental processes. Complex thinking is used with spreadsheets to analyze, reason, and construct simulations. In some ways, Jonassen's mindtools are simply an extension of the theory that supports computer programming—they both involve simple computer programming that focuses the learner on the process of calculations and other simple functions. Jonassen's contribution is that he points out that programming can be done in simple ways with widely available software.

CONCLUSION

In this chapter, we discussed a group of specific tools for learning in computer environments using various cognitive models. We saw that the notion of critical thinking is not very well defined in the research literature and therefore elusive as a goal of teaching. Concept maps, knowledge modeling, and computer programming were examined as specific techniques for encouraging deeper levels of thinking and learning by focusing on process. Finally, cognitive tools that can amplify thinking and learning were discussed. In the next chapter, the final of the first part of this book centering on computer-based theory and practice, I turn to important faculty and teaching issues uncovered both in the data collected and in the research literature.

Chapter VIII

Faculty and Teaching Issues

The technological is allowing me to reexamine things from a different angle (Don Cardinal, Chapman University, School of Education Faculty member).

I would be remiss if I ended the first part of this book on the theory and practice of computer-based computer environments without addressing teaching issues. Faculty and teaching issues repeatedly came up in both the 2000 and 2001 studies. Clearly, teaching and learning issues affecting both administrators and students are of central concern when considering the pedagogy of computer-based educational programs.

One of the first questions asked about distance learning is how does its effectiveness compare to that of traditional methods? The 2001 survey found that respondents tended to see distance learning as an effective method for learning with 86.9% of those in computer-based courses strongly agreeing or agreeing with the statement, "I learned as much or more in this distance learning course as in an average traditional face-to-face course" (See Figure 19).

Figure 19: Learning Comparison by Delivery Format (Questions 1 & 2)

Delivery format * I learned as much or more in this distance learning course as in an average traditional face-to-face course. Crosstabulation

			strongly agree	agree	disagree	strongly disagree	Total
Delivery format	computer-based	Count	23	30	6	2	61
		% within Delivery format	37.7%	49.2%	9.8%	3.3%	100.0%
	videotape	Count	10	30	16	6	62
		% within Delivery format	16.1%	48.4%	25.8%	9.7%	100.0%
	correspondence	Count		3	1	1	5
		% within Delivery format		60.0%	20.0%	20.0%	100.0%
	other	Count		1			1
		% within Delivery format		100.0%			100.0%
Total		Count	33	64	23	9	129
		% within Delivery format	25.6%	49.6%	17.8%	7.0%	100.0%

Note here that the videotape course respondents were less enthusiastic, with only 16.1% strongly agreeing with the statement, as opposed to 37.7% of those taking computer-based courses. Although the data collected for this survey is too slight in number to be conclusive, it indicates that computer-based courses are more effective than video-based ones.

To consider this comparison question a little more deeply, respondents were queried about the level of critical thinking used in the courses. The 2001 survey found that students in computer-based courses felt that critical thinking skills (not defined) had been used, with 93.4% either strongly agreeing or agreeing with the statement, "Critical thinking skills were utilized and developed in this course."

Figure 20: Critical Thinking Utilization by Delivery Format (Questions 1 & 5)

Delivery format * Critical thinking skills were utilized and developed in this course. Crosstabulation

			strongly agree	agree	disagree	strongly disagree	Total
Delivery format	computer-based	Count	28	28	3	1	60
		% within Delivery format	46.7%	46.7%	5.0%	1.7%	100.0%
	videotape	Count	12	35	12	3	62
		% within Delivery format	19.4%	56.5%	19.4%	4.8%	100.0%
	correspondence	Count	2	2	1		5
		% within Delivery format	40.0%	40.0%	20.0%		100.0%
	other	Count	1				1
		% within Delivery format	100.0%				100.0%
Total		Count	43	65	16	4	128
		% within Delivery format	33.6%	50.8%	12.5%	3.1%	100.0%

Note here, as with many of the responses in the 2001 survey, that the students in the videotape courses reported a lower degree of agreement with this statement (19.7% "strong agree" and 55.7% "agree").

In terms of teaching through memorization and repetition in distance learning courses, students in computer-based courses indicated that this was not a primary method of teaching, with 85.0% disagreeing or strongly disagreeing with the statement "I primarily learned in this course through memorization and repetition," as seen in Figure 21.

Figure 21: Learning Through Memorization and Repetition by Delivery Format (Questions 1 & 6)

Delivery format * I primarily learned in this course through memorization and repetition. Crosstabulation

			I primarily learned in this course through memorization and repetition.				Total
			strongly agree	agree	disagree	strongly disagree	
Delivery format	computer-based	Count	1	8	30	21	60
		% within Delivery format	1.7%	13.3%	50.0%	35.0%	100.0%
	videotape	Count	9	29	22	1	61
		% within Delivery format	14.8%	47.5%	36.1%	1.6%	100.0%
	correspondence	Count		1	3	1	5
		% within Delivery format		20.0%	60.0%	20.0%	100.0%
	other	Count		1			1
		% within Delivery format		100.0%			100.0%
Total		Count	10	39	55	23	127
		% within Delivery format	7.9%	30.7%	43.3%	18.1%	100.0%

Note here again that videotape course respondents responded at a lower rate of disagreement to this question, with only 1.6% strongly disagreeing.

In the interviews conducted for the 2000 survey, administrators consistently spoke about how the implementation of distance learning led to a broader consideration of teaching/learning as a whole.

> Faculty seem to have the right to not be critical about their own traditional teaching. But when you teach online you've lost that right, and now everyone needs to be critical about their own teaching. You can go back into the classroom and be a crappy teacher, and so you receive a letter once in a while—if you are tenured, nobody can do anything about it....So I think that it's allowing us to take some things out of the closet that have been there for a long time....So I think these are the artifacts, the by-products of online courses. It's not so much that technology is teaching, but that I'm getting to look at teaching in a different way (Don Cardinal, Chapman University, School of Education Faculty member).

One respondent spoke about how the scheduling of face-to-face meetings was being reconsidered in conjunction with the technologically augmented classroom.

> The important thing is that people are sitting down and evaluating their objectives. If one instructor thinks that they should meet more, I think that's legitimate (Warren Ashley, Director of Distance Learning, CSU Dominguez Hills).

Another respondent spoke about the controversial issue of how to measure seat-time in a distance learning course.

> Can they learn more in a shorter amount of time?....When I'm in a class I have the students go out and collect data. Is that work the three instructional hours? That's the way I look at it (Don Cardinal, Chapman University, School of Education Faculty member).

Many respondents spoke about how teaching in computer-based environments is more difficult than in face-to-face environments.

> You have to know what you are doing. A lot of times, in a traditional classroom, an instructor can fake it. All they have to do is go in and start talking about something that is related to the theme of the course, and it works. You can't do that in distance learning.
> Not only do they have to know exactly what they want to communicate, they have to have it planned so that there's no question in the mind of the students about what is being communicated. Any ambiguity will create confusion. There's much more planning and decision making required of distance learning faculty. We need to have the outline for the course at least a week. For most faculty this is incredible, they've never done this. The good thing is that if you do it well the first time, chances are you won't need to make a lot of revisions the second or third time (Warren Ashley, Director of Distance Learning, CSU Dominguez Hills).

One respondent spoke about how the successful use of email has led to a greater appreciation for the need of interaction with students.

> I think email is a big one. Why is it that when I have online assignments, that I get an average of twenty emails per student per month? Why do

I get to know those people better, and we're closer, and the learning increases? Well, I think it's just contact with those people, and it should generalize to the rest of your teaching (Don Cardinal, Chapman University, School of Education Faculty member).

Furthermore, this respondent found that technology has helped to gain an understanding of a less hierarchical classroom management style.

Maybe being more collegial with students, not so top down with students. After twenty years of teaching in higher education, that's been my gift from online learning (Don Cardinal, Chapman University, School of Education Faculty member).

Distance learning has also led to this faculty member considering how support materials such as textbooks can most effectively be used.

I think the struggle to use technology and understand the pedagogy is useful for them. Just like deciding what books to use, what materials to make available. Right now I'm trying to decide to use an online book, or a paper book. The issues are similar to deciding which text to use. It's a way to make new, old issues that we take for granted (Don Cardinal, Chapman University, School of Education Faculty member).

This faculty member also pointed out how experience with distance learning has led to methods for increasing course content and longer office hours.

When you get this increase you can't necessarily connect it to the technology, maybe it's my own awakening to pedagogy. Let's increase office hours, so now you don't know what to attribute it to. Why have I been able to add curriculum to a course that was already notorious for being packed? (Don Cardinal, Chapman University, School of Education Faculty member)

Contrary to all of these positive comments about distance learning's affect on pedagogy, one respondent using videoconferencing technology argued that technology probably hasn't changed teaching methods much.

I'm sure it varies from professor to professor, but it probably varies in the same way even if they weren't in our classroom in the sense that if they use certain kind of visuals, they probably would have done that in any case (Program Manager, anonymous large, independent, Western U.S. doctoral degree-granting institution).

One administrator from a religious-based institution pointed out the problem with motivating students in distance learning courses.

The second issue is accountability, the distance ed[ucation] courses are not as urgent, easier to kick under the bed; they tend to not get as much attention as in-person courses, when sometimes students take both at the same time. We have some students doing that because of scheduling conflicts, or whatever; they tend to pay more attention to the in-person courses because they seem more urgent. With a job, a ministry, a life, it is easy to put these off (Jon Raibley, Assistant Director of the Center for Lifelong Learning, Western Seminary).

This same respondent indicates that in order to meet this motivation problem the difficulty of the workload has been increased.

So, with the new courses we are trying to come up with a more rigorous schedule. The older courses have been on an independent study basis (Jon Raibley, Assistant Director of the Center for Lifelong Learning, Western Seminary).

This is a common experience reported in the 2001 survey where many students indicated that distance learning courses were more difficult.

This administrator from the previously cited religious institution spoke also about the need for some face-to-face to assess character development. This is an important question for distance learning as a whole. Can deeper character and personality development occur at a distance using technology?

I think what we are doing is a little special because of who we are targeting and what we are aiming to do. I studied leadership issues in the church, how have leaders been trained? For us we either focus on spiritual knowledge, academic knowledge, or practical skills. The church has found academic knowledge works well with distance ed[ucation]. Our challenge is to not stop there. Because there are skills, interpersonal skills, etc., there are other ethical issues—we can't have pastors dealing with character issues if we haven't had contact with them....With the new courses we've talked about the strengths of distance ed[ucation], and some of the faculty are quick to point out the weaknesses. They can't assess character; they can't assess skills. In some cases that means bringing in guest speakers, speakers with particular skills in an area, use media pieces that we wouldn't use in an on-campus course (Jon Raibley, Assistant Director of the Center for Lifelong Learning, Western Seminary).

One strategy shared by this religious institution is to use on-site mentors.

These courses will be done with on an on-site mentor who will be helping us assess the character and skill development. That's an important part. It's a team effort....Because while we can provide the information, there's more to what we are trying to accomplish than just transmitting information (Jon Raibley, Assistant Director of the Center for Lifelong Learning, Western Seminary).

One respondent spoke about the difficulty of relying just on email for distance learning courses.

When we first started offering courses online, we used only email. Instructors found offering courses via email too much work, inefficient, and ineffective; they were overwhelmed with the volume of student email. (Vivian Sinou, Dean, Distance & Mediated Learning, Foothill College).

One respondent to the 2000 survey indicated the use of a blended approach of face-to-face and online courses work, which is becoming increasingly popular.

This new course we are starting next semester will be a required fifty percent online, and all the students in the master's program will have to take it, and they will have to participate in a group area. They will have to participate in some chat areas to see what that is like (Don Cardinal, Chapman University, School of Education Faculty member).

The 2001 survey gives us the student perspective on teaching/learning in distance learning. We see here that lack of support from faculty members seemed to be much more obvious for students in distance learning courses than in traditional courses.

Instructor makes very little effort to meet with students and she never returned any of my work....I never saw her once, she was always 'too busy' (anonymous student in distance learning course).

> I don't feel I have been given the right amount of attention pertaining to questions I have had. I feel simple answers should be received in a timely manner–not having to call back a week later (anonymous distance learning student).

> I found getting in touch with the teacher and getting sound feedback was next to impossible (anonymous distance learning student).

On the other hand, when faculty members were responsive in distance learning courses, they were effective.

> I found the instructor to be instrumental in my online course. He was extremely helpful, knowledgeable, considerate and very fast in communicating a response to me either via email or the message board (anonymous distance learning student).

The overall difficulty of distance learning courses was commented on time and again; as we saw earlier, in order to address the academic concern for seat time, a large amount of content is often added to compensate for this perceived deficiency.

> If given the opportunity again to take a telecourse, I would decline. There was too much information to take in (anonymous distance learning student).

> I feel that independent study courses have always required more work from me. I had to organize, manage, and use time well. The amount of work was large to make up for the lost lecture and interaction with instructor (anonymous distance learning student).

> While my [video]tape course served its purpose, I did not enjoy it, nor do I believe I learned much. The class attempted to make up for lack of interaction with a bombardment of material. It was self-defeating (anonymous distance learning student).

Additionally, students felt that the grading scale for distance learning courses was more difficult.

> This is a crazy teacher for the way she grades (anonymous distance learning student).

> I pulled my lowest grade [in this class], and do much better in a classroom setting (anonymous distance learning student).

Despite specific complaints mostly about faculty members lack of attentiveness, students in distance learning courses report that their experience was an effective way to learn. The data show that computer-based courses compare favorably with face-to-face courses in terms of learning outcomes, use of critical thinking, and an avoidance of learning primarily by memorization. Faculty members reported that teaching distance learning courses has led to a reexamination of teaching, having a positive effect on their face-to-face courses. The use of office hours, support materials, group projects, and individual communication with the students is reconsidered often through teaching distance learning courses. However, some faculty reported that teaching distance learning format courses was more difficult. Likewise, students reported that often the distance learning courses were loaded with content to compensate for the lack of face-to-face time. Faculty members also commented on the difficulty of more subtle forms of communication such as judging student character in online environments. For students, it is clear from the data that the number and quality of the interactions with the faculty members is a key to the success of the learning experience.

Finally, one main way faculty work is changing is through the use of the Internet for research. In the 2000 survey, a faculty member commented on the way that the Internet has opened communication for students for research opportunities by enabling communication with scholars in their field.

> Already in the last two years of doing this, students will call authors of articles. Before, they would never do that. Now it is quite common. The world seems to seem smaller to them, so much is accessible, why not the authors of their readings? It's easier to get phone numbers off the Web, email addresses, and this has given them a whole different understanding of research—they're just people. They call authors and the person says, 'You know, tonight's Halloween and I'm taking my kids out.' They look at you and say, 'God, Don, they have kids. I have kids, they have kids, we're the same.' So there is an equalizing aspect (Don Cardinal, Chapman University, School of Education Faculty member).

This same openness to scholars in the field has clearly affected the research work of faculty. Currently there is a shift from paper to digital media in the submission, review, and publication of scholarly work. Increasingly, established journals are offered in both paper and online versions, and the submission process for peer review has been substantially sped up by the use of the Internet. One obvious need in the future is for software developed to orchestrate scholarly commentary and exchange of information before publication.

This concludes the first part of my book occupied with an examination of the current practice and theory of distance learning. Much of what has been reviewed has been discussed and debated in various forums by other scholars. The data from the two original surveys helped to make the discussion concrete and up-to-date. In the second part, I turn to an exploration of new territory for distance learning by envisioning the computer as a medium. Additionally, I look at theories that might be used in constructing computer educational environments, visions of future learning devices, as well as charting a research agenda for the future.

Part II

Computer as Medium

Chapter IX

Film Theory

The second part of this book investigates the computer learning environment as a medium by looking at media theory and relevant film criticism, including documentary and fiction film theory. Media theory reveals the social aspects of the development of new media and places computer environments within a larger tradition. Recent film theorists who use cognitive approaches to explore meaning construction within a viewer also provide a foundation for understanding the computer user. Film narrative and non-narrative conventions help illuminate key issues in computer environments' use of narrative. Specific applications of theory in the use of simulation as well as case studies are discussed. Finally, I conclude with a review of what we know about the design of computer learning environments, and offer some speculative thoughts on the future uses of technology in education. As opposed to the first part of this book that detailed what we know about distance learning, this second half is intended to spark ideas for designers and educators about new ways to learn.

CURRENT PRACTICES

In this chapter, I review the survey findings on current use of media in education, review film and media theory, and consider the specific ramifications of computer as a medium used for education. I begin with the current use of media in American distance learning format courses. As is true of most things in higher education, researchers seeking to understand current practices are immediately confronted with a rich variety of institutional practice. Lynch (1998) describes the experience of George Washington University, an early leader in distance learning in the United States, paralleling the experience at many institutions over the last decade where the evolving technologies led to a constant upgrading and transformation of delivery systems, moving from low to high quality video, then to web-based and integrated digital video. News talk or variety-style video productions came to George Washington University with the collaboration of a for-profit company, Jones Entertainment, with camera movement, three-camera style, and scripted broadcast-quality techniques the norm in this transition. Lynch further describes how the move to the Web led to a rethinking of video as only one content source, no longer as the delivery system itself. Currently, the inability of digital technology to accommodate large video files has thus far limited their use in distance learning. The stress of large image files has led to small video screen sizes, with medium shots and close-ups used with short takes to compensate for the necessary technical limitation.

While undoubtedly computers are becoming better able to accommodate video, this experience of evolving style coupled with quickly changing technology was a common occurrence in many higher education institutions over the past decade. In addition to the evolving technology, management issues have affected how media are used in distance learning courses. Lynch relates how George Washington University ran into problems with scheduling in the tight format in terms of the course content and in working with faculty unaccustomed to a demanding production schedule. He claims the reduction in the reliance on video created a more interactive and integrated learning environment for students in the early stages of computer-based learning.

Partly because of the technological limitations, the George Washington University history reveals, and because of a lack of understanding about how and when to use media, most distance learning courses are media poor. In the 2000 survey, administrators were asked to compare the experience of taking one of their distance learning courses to another medium. In response to the question of which of the following is most like taking one of their distance learning courses, they responded that the experience most closely paralleled reading a book (27.3%) or writing letters (25.5%) (see Figure 22).

Figure 22: Comparison with Other Media (Question 46)

	Reading a book	Watching TV	Watching a movie	Listening to the radio	Talking on the telephone	Writing letters
Percent	27.3%	21.6%	5.0%	4.2%	16.3%	25.5%
Number	77	61	14	12	46	72

question of which of the following is most like taking one of their distance learning courses, they responded that the experience most closely paralleled reading a book (27.3%) or writing letters (25.5%) (see Figure 22).

These data reveal that those responsible for developing distance learning courses are still primarily focused on text-based approaches. Watching television was the third most common response (21.6%), which undoubtedly has to do with the large number of institutions involved in the use of telecourses. Although television was the third highest choice, much of instructional television relies heavily on taped lectures and is not very sophisticated or media rich. Notice that respondents felt that their courses were least like watching a movie (5.0%) or listening to the radio (4.2%). These data suggest that sophisticated approaches to the design of distance learning courses using techniques from mature media such as film are not generally utilized.

Nevertheless, students seem to appreciate media-rich learning environ-

Figure 23: Media Added to Learning by Delivery Format (Questions 2 & 7)

Delivery format * The media used in this course added significantly to the learning experience. Crosstabulation

				The media used in this course added significantly to the learning experience.				Total
				strongly agree	agree	disagree	strongly disagree	
Delivery format	computer-based	Count		31	27	1	1	60
		% within Delivery format		51.7%	45.0%	1.7%	1.7%	100.0%
	videotape	Count		11	30	16	5	62
		% within Delivery format		17.7%	48.4%	25.8%	8.1%	100.0%
	correspondence	Count		1	3		1	5
		% within Delivery format		20.0%	60.0%		20.0%	100.0%
	other	Count			1			1
		% within Delivery format			100.0%			100.0%
Total		Count		43	61	17	7	128
		% within Delivery format		33.6%	47.7%	13.3%	5.5%	100.0%

Oddly enough, the videotape course students responded at a lower rate of agreement, with only 17.7% "strongly agree" compared to 51.7% for computer-based students. Perhaps this is a comment on the effectiveness of the use of media in telecourses. Regardless, we can see from the combined data of the two surveys that while students confirm the importance of rich media, the courses offered still tend to be largely text-based efforts.

The interviews provide a more detailed look at some of the media practices of those leading institutions most involved in distance learning. Those active in telecourse video production spoke about the attempt to make broadcast-quality productions.

> We've always thought that the quality needs to be there, not just the academic quality but the production quality as well....Pretty much use a broadcast approach (Vice-President, anonymous large, independent, Eastern U.S. doctoral degree-granting institution).

The following excerpt tells how the hiring of staff with commercial television experience led naturally to this approach.

> My feeling was that we needed to look very professional. So we created more of a set than what you usually see when people do instructional television. The technology goes way back. We used an Amiga computer for graphics which was more legible than the instructor writing on the board or using an overhead camera. Then as we grew and as I was allowed to hire staff (because at first there was just an engineer and myself), we hired people with TV experience. Of course, in LA [Los Angeles], it's hard to find someone who doesn't have TV experience. I never put that down, it was never one of the criteria that we looked for, but my senior producer has over 200 commercials to his credit, and my broadcast operations coordinator worked for the Home Shopping Network—so I have many people with professional TV experience. And they bought into the idea that we should look as professional as possible. They were always on the lookout for ways that we could improve our signal, our broadcast, our look. And there were a lot of things that we did that other campuses didn't do. This was largely because of my staff (Warren Ashley, Director of Distance Learning, CSU Dominguez Hills).

The quest for broadcast quality video led to the use of better lighting, graphics, and sets.

> One of things that anyone familiar with TV knows, is that the more lights you have the better the picture. So we put in lots of lights. We had three or four times as many lights as other campuses used. And it gives you a much better picture and a sharper look. We eventually figured out a way to put graphics up on a monitor in the background, and although we knew we weren't going to look like the networks, we were modeling ourselves after CNN (Warren Ashley, Director of Distance Learning, CSU Dominguez Hills).

In relationship to the use of video in digital computer platforms, respondents also spoke of lessons learned.

> We've learned a lot. For instance, broadcast video doesn't work well in streaming video. The kinds of backgrounds you use make a big difference. If you have someone outside standing in front of a tree with the wind blowing, that takes up an incredible amount of bandwidth. If you put the person in front of building instead, you save yourself bandwidth. In broadcast situations you follow walking figures with the camera, but for streaming video you want to try to arrange it so you don't have to follow the person and can then save a lot of bandwidth (Vice-President, anonymous large, independent, Eastern U. S. doctoral degree-granting institution).

In the interviews, administrators spoke about some of the visual techniques they use. One interviewee spoke about the use of lecturing or "talking heads" in video and computer-based video.

> Some people are wonderful lecturers and are enthralling. Others just lecture and it's boring. That's part of the role of the instructional designer and production people to figure out how best to present the content. We have a cooperative program coming up this spring with Texas A & M, and the older George Bush will be presenting. I think we will want to listen to what he says. But the old idea of turning a camera on in a lecture classroom we found didn't work twenty years ago. Yet you still see it (Vice-President, anonymous large, independent, Eastern U. S. doctoral degree-granting institution).

Respondents spoke about how the use of the Internet in distance learning has changed the way video is used. When the video isn't used to record a whole lecture but only to supplement other forms of delivery, it takes on a specific use.

I have begun suggesting to the faculty who are using the Internet in a more substantive manner maybe they don't need to broadcast a three-hour class every week. Maybe some weeks they need three hours to meet their objectives but other weeks they don't. And maybe some weeks they can do all of the instruction in asynchronous manner. This is a difficult concept for many of them because they have always taught the same number of hours each week—now they have to decide what is needed for each class (Warren Ashley, Director of Distance Learning, CSU Dominguez Hills).

Here we see the faculty looking more closely at when to use video—a result of online communication delivering an increasingly large part of the course. One technique consistently used in educational media is to show concepts introduced through the use of video.

Whenever a new concept is introduced it is shown being implemented in the classroom…footage will show a teacher in the classroom implementing something in an actual lesson (Joy Edwards, Director of Graduate Studies, Texas Wesleyan University).

In a program in which I teach, we use video to show how something works, or to show a piece of equipment, or to take a field trip, or to interview a guest, but the lecture is mostly textual. We have graphics and animation built into it (Vice-President, anonymous large, independent, Eastern U. S. doctoral degree-granting institution).

Confirming what this interview subject said, the 2001 survey respondents strongly indicated (91.5% "strongly agree" and "agree") that video was useful in demonstrating or showing content.

Figure 24: Video Use in Showing Content (Questions 1 & 12)

Delivery format * Video is useful in showing or demonstrating content. Crosstabulation

				Video is useful in showing or demonstrating content.				Total
				strongly agree	agree	disagree	strongly disagree	
Delivery format	computer-based		Count	7	33	2	1	43
			% within Delivery format	16.3%	76.7%	4.7%	2.3%	100.0%
	videotape		Count	12	31	3	1	47
			% within Delivery format	25.5%	66.0%	6.4%	2.1%	100.0%
	correspondence		Count		2		1	3
			% within Delivery format		66.7%		33.3%	100.0%
	other		Count		1			1
			% within Delivery format		100.0%			100.0%
Total			Count	19	67	5	3	94
			% within Delivery format	20.2%	71.3%	5.3%	3.2%	100.0%

We see here that students in both computer-based and video courses see the value in the use of at least some video to present content.

The data from both administrators and students indicate that we are in an early stage of development in the use of media in education generally, let alone in computer-based educational environments. In reviewing the material, one is struck by the variety of approaches and the quickly evolving use of technologies. Furthermore, unlike classroom instruction that has generated a plethora of books on techniques and theory, the use of media does not yet have strong theoretical and scholarly support. Consequently, administrators are still grappling with this new medium's learning issues on a trial and error basis. To begin to develop stronger theoretical support, what might we learn from disciplines with more developed use of media?

HISTORY OF FILM

In order to understand educational software as a new medium, it is valuable to have a historical perspective of the development of other related media,

Figure 25: Story Slide—Miss Jerry (1894)

especially film. What has been the positioning of the audience to film projections? What have been the educational and ideological purposes and uses of film?

From the beginning, film has had a magical ability to imitate reality in a convincing fashion, while at the same time it is a manipulation of the human mind. The roots of the film medium can be traced to the magic lantern that originated around 1660. In what was known as the "phantasmagoria" in traveling tent shows, shadows were projected from behind on a make-shift screen.

This technique continued to evolve until the 1880s, when "story slides" became popular. Story slides were photographs using actors with props in sets projected on a screen from behind. A system of lanterns was used to provide transitions from one slide to the next, even dissolving and superimposing two images together, as well as fading into one another—a clear precursor to film transitions. Brownlow and Kobal (1979) argue that early silent films overall were very similar to these story-slides. This combination of reality and unreality seen in the story slides presented at traveling tent shows has continued to be an essential appeal for film audiences, and now computer users. Although these early story slides provided entertainment and were a technological curiosity, soon their potential use for persuasion was understood.

IDEOLOGY AND FILM

To gain a full understanding of the implications of the use of computers in educational environment, one needs to recognize how ideology is transmitted through a medium. Early on, film was recognized for its propaganda potential. In 1917, Lenin proclaimed cinema as the "art of the people" and the Soviets understood film's potential propaganda power. Silent film in America had a social orientation that was aimed more towards a mainstream audience than the Soviet's rather intellectual early filmmakers (Mitry, 1997). According to Brownlow (1968), motion pictures were "the common language of the poor" in America. He sees the success of early American filmmakers rooted in their lack of formal education or a broad knowledge of literature; consequently, they relied more on visual narrative methods to communicate with their like-minded audience. This appeal to the masses, recognized by both Lenin and early American filmmakers, lent the medium to the easy transmission of political viewpoints.

In the period of the 1920s through the 1950s, it was quite common for writing and directing talent to come from diverse walks of life, and often resulted in Hollywood films that were simple and straightforward in their style. It wasn't until the 1960s that academically trained talent began to work in the film industry. In looking at the broad appeal and accessibility of early Hollywood film, one can't help but wonder if, through the effective use of this new computer medium, distance learning might also become the common educational language of a new generation of working-class Americans who have not traditionally had access to education. We will see later how the extensive theoretical literature on both fiction and non-fiction film looks very closely at ideology. At this point it is important to note that this ideological quality of the medium was appreciated early on in its development.

ACADEMIC LITERATURE ON FILM AND EDUCATION

For the most part, educational technology has not made use of the extensive film criticism literature. Ellsworth and Whatley (1990) note that educational technology is preoccupied as a field with questions of how viewers learn through media and the isolated effects of specific media techniques. As opposed to the film theory literature, ideological questions centering on meaning and significance, or ideological issues, are not often addressed in educational technology. Often in educational technology literature, the transmission of knowledge through media is seen as neutral—this is far from the case. On the other hand, film criticism focuses on questions of authorship, psychological, feminist, and semiological approaches to media, and ignores less enthralling educational media topics.

Nevertheless, in the past few years some critics have begun to look at the connection between media and education. Ellsworth and Whatley (1990) argue that key ideological questions include: How does the use of visual representation in curriculum materials elevate some ways of knowing over others? How do the terms and interest of such privileging relate to the school's role in society? A sociology of education points to knowledge as being socially constructed both in methods and forms of conveying knowledge. One critical view holds that schools organize curriculums in ways that position students as consumers, passively taking in goods. According to Ellsworth and Whatley, ideological analysis looks at absences in media texts that reveal bias and the author's point of view. One response is to define these absences in order to restructure the relationship of learners to media and make the ideological analysis of educational media a strategic practice used regularly in classrooms.

Ellsworth (1990) argues that conventions of visual representation and narrative form found in most educational films contradict the goals of pedagogical strategies to make students active producers of meaning. He notes that instructional films often invite viewers to take up particular physical, social, and ideological involvements or points of view. We will see later in this book the importance of subjective narrative positions in media. Ellsworth claims that the viewing experience creates relations between an individual's knowledge and position of power in society. In this way, it is common in educational films to position the viewer outside of the film world looking in—a powerless positioning. While this positioning may be simply an attempt at scientific objectivity, the film medium emphasizes the removal found in the lecture hall and increases the effect.

Instructional films often use before/after and problem/solution narrative structures that imply viewer ignorance before viewing, and then knowledge after seeing the film. This positioning leads the viewer to desire to possess the narrator's knowledge. The student needs the narrator's after story sequence or ordering of knowledge. The series of shots in these films often break objects down into parts for clarification. The use of a narrator positions the viewer, both through the camera position and voice-over narration, in a subservient location. Often white and male, the narrator is privileged with knowledge and has what might be described as a paternal positioning. In this way, Ellsworth (1990) claims that the viewer moves from child to adult in the progress of the typical instructional film narrative, beginning with ignorance and ending with grown-up knowledge. He argues that this style has changed somewhat since the 1960s, when techniques of cinéma vérité and television documentaries emerged in instructional films, which included modifying and sometimes rejecting the male, paternalistic narrator.

Erdman (1990) argues that originally instructional films drew from popular film for stylistic conventions. As with popular films, instructional films are self-contained systems. Animation and cinematic effects are common in instructional films, as are close-ups and extreme close-ups. Static framing and close angles keep viewers from broadening their attention. Controlled viewing is common in instructional films, as it is in Hollywood films. Fades and dissolves are used to indicate changes in emphasis. Also, fade-ins and -outs are used to mark changes between sections of the instructional films. Thus one can see that instructional film does to a large extent borrow from traditional Hollywood film stylistically.

Using a notion of film style as defined by repeated use of techniques for specific effect, Erdman points out the differences between traditional classroom lessons and the structure of instructional films. Traditional lessons include stated learning objectives, recalling prior relevant student learning, motivating, presenting content, reviewing, applying knowledge, and evaluating. Conversely, instructional films are structured very simply with an introduction or prologue, presentation of content through demonstration and application, and a summary. Most importantly, the instructional film does not refer to students' prior experiences—there is no effort to customize the learning experience or to motivate students. Although there are occasionally a few sentences referring to student experiences, most instructional films are structured to provide motivation purely from the content. So we see that instructional films generally have adopted Hollywood film techniques and have thereby often neglected standard classroom strategies for involving students.

Undoubtedly, cost of film production has limited the ability to customize the viewing experience when addressing a diverse and large audience.

We see here that the limited literature of educational technology that looks at media properties raises many important issues that I herein examine in this second part of the book, including the closed, unsophisticated use of film, lack of interactivity or personalization of approach, and the audience's submissive positioning.

COMPUTERS AS MEDIUM

Computers are usually viewed as tools or instruments for storing and manipulating data. However, at times the literature on human-computer interaction (HCI) suggests that the computer is a medium, not a tool, and that further investigation might prove fruitful (Baecker & Small, 1995; Head, 1999; Kay, 1995; Oren, 1995). As the use of computers in educational environments increases, a more sophisticated understanding of computer design issues becomes more important—a revisioning of computers as a medium brings this kind of complexity to the research.

The research literature suggests that the evolution of computers might parallel other forms of media (Mountford, 1995; Oren, 1995). Marshall McLuhan (1964) pointed out that new media are dependent on old media until the unique features of the new media are appreciated and developed. In the way that early movies relied on novels and plays for content, early computing automated the work of typewriters and accounting ledgers. Software engineering has been dominated by engineers; similarly, the same profession first controlled the development of the filmmaking process. Because engineers rather than forward thinking educators have controlled the process, much of the current educational software simply automates typical classroom tasks. The understanding of computers as a medium may be a key to re-envisioning educational software. Now that the personal computer is reaching a more mature stage in its developmental cycle, it is time to look more closely at the specific characteristics of the medium for educational applications.

Oren (1995) argues that understanding computers as a medium means enlarging human-computer interaction research to include issues such as the psychology of media, evolution of genre and form, and the societal implications of media. Computers began to be used in educational environments much later than film, and we are still using computers instructionally at very unsophisticated levels. This chapter began by emphasizing this point by citing data on current practices, which reveal high use of text over other media. If computers are a new medium, what is unique about it? Certainly, the use of standard text and recorded lectures is not new. What are the computer's specific advantages?

Gibbons and Fairweather (1998) identify five attributes that make the computer unique as an instructional medium: a dynamic display, the ability to accept student input, speed, ability to select, and flawless memory. Taking the last point first, one of the distinct advantages of learning in computer environments might be this ability to have a perfect record of learning. Plaisant, Rose, Rubloff, Salter, and Shneiderman (1999) state that such a record could help students monitor their behavior, reflect on their progress, and experiment with revisions of their experiences. Providing interactivity through selection and student input with speed comprise obvious attributes discussed through this book. In terms of a dynamic display, some argue that computer environments are particularly useful in giving students rich learning experiences, a direct result of its media nature. Shaffer and Resnick (1999) describe a "thick authenticity" in computer simulation that is personally meaningful and connected to the real world. We see later in this chapter how this authenticity comes largely through the use of photographic images, but first I want to broaden the discussion of computer as medium with an overview of communications and media theory.

COMMUNICATIONS AND MEDIA THEORY

Communications theory dates back to Plato and Aristotle, but became a recognized academic field in the 1940s during World War II (Newby, Stepich, Lehman & Russell, 1996). Media theorists have analyzed the development of new media and how they are connected to broader social evolution. These theories are useful to examine in regard to understanding computer learning environments as a new medium because they place it in a larger historical context. Often in discussions about distance learning it is easy to focus on the immediate connected issues and lose sight of the long-term development of new media and technology.

A tradition of scholarship focusing on communications effects led to research on the media industry, the military, and functionalism. Many concentrated simply on the short-term effects of media (Klapper, 1960). However, others looked at it more broadly, particularly in terms the transmission of ideology. An important influence on media theory was the Department of Sociology at the University of Chicago in the 1920s and 1930s, where a theory showing the importance of communication in social life using ethnographic methods to explore complex social interactions emerged.

Various scholars, including Lewis Mumford and Harold Innis, commented on the evolution of media and analyzed how they connect to other historical developments. In *Technics and Civilization*, Mumford (1934) argues that the printed sheet was the first completely mechanical achievement in the West. Writing was a great labor-saving device over oral communication and released people from a dependence on the present time for communication. Innis (1972; 1991) argues that studying forms of communication offers possibilities to understanding government and the rise and fall of empires in Western civilization. Changes in political form coincide with adoption of new media. Thus the shift in Egyptian civilization from a monarch to a more democratic organization coincided with shift in emphasis from stone to papyrus. The properties of the dominant medium, along with institutional structure, facilitate knowledge and power. According to Innis, the Roman Empire involved the destruction of the oral tradition and the imposition of writing.

Education has been dominated by the written medium since the rise of the Roman Empire. Does the rise of the computer medium mean a shift in power and perhaps finally a shift away from the dominance of the print medium on education? Innis' solution is to promote the use of dialogue as a form of communication. The oral tradition is important because it emphasizes dialogue and works against monopolies of knowledge. Aware of John Dewey and Robert Park's work in the Chicago School, he argued for an oral tradition revival in education. A central force in educational theory, John Dewey in *The Public and Its Problems* (1927) saw communication media as an extension of a free press with the purpose of disseminating freedom. Innis claimed that the conditions of freedom of thought are in danger of being destroyed by technology and the mechanization of knowledge. He saw one of the aims of education as breaking the strong hold of the present on the mind. Innis held that to reach the lower levels of intelligence and to concentrate on territory held by newspapers and radio, adult education follows the methods of commercial advertising. One could argue that the computer medium with its interactive communication capability is closer to dialogue than other media such as print, film, or radio—but I suspect Innis would not see it that way.

According to Meyrowitz (1986), electronic media bring back a key aspect of oral societies: simultaneity of action, perception, and reaction. In this way, sensory experience again becomes a prime form of communicating. Yet electronic media is far different from oral communication because it is not subject to physical limitations of time or space. Print allows for new ways of sharing knowledge, while electronic media tend to foster new types of shared experience. Meyrowitz views television and other electronic media as breaking the age-old connection between where we are and what we know or experience. According to Meyrowitz, McLuhan saw the problems of the traditional school as linked to the shift from a mechanical age characterized by fragmentation, specialization, and "meanness" to an electronic age characterized by wholeness, diversity, and deep involvement.

Eisenstein (1979) argues that the preservation of knowledge was the most important aspect of printing. The notion that information is best preserved by being public not private was a key element of printing as an advancement in civilization and ran counter to the previous secretive nature of knowledge. Additionally, Eisenstein sees the increasing familiarity with numbered pages, punctuation marks, sections breaks, running heads, and indices as helping to reorder the thought of all readers. She argues that a reading public is more individualistic than a hearing one. Eisenstein points out that both religious and technical texts spread widely with the rise of the printing press. Consequently, both religious and scientific traditions were greatly affected by the advent of printing. In opposition to Mumford, Innis, and McLuhan, she points to the difficulty in generalizing about the consequences of media advancements; according to Eisenstein, the effects of changes in media are not as clear cut as others have argued.

Sholle and Denski (1994) argue that mass media represent the greatest force for social control ever imagined, and media education presents mechanisms and techniques for control. They propose that the bridging of the theoretical traditions of media studies with critical pedagogy may provide a solution to a divided approach to media studies. Critical pedagogy focuses on political/economic issues of schooling such as representation of texts and construction of subjective states of mind in the student. Critical pedagogy defines school as cultural politics, as a way of maintaining or modifying discourse, and appropriating knowledge and power. Critical pedagogy of media begins with an assessment of contemporary culture and the function of media within it. Sholle and Denski claim that we need to understand how media affect everyday life and to help students become media literate, to understand multiple references and the codes that position them as learners.

In his influential book, *Ways of Seeing* (1972), John Berger argues that the modern means of reproduction have destroyed the authority of art. Photography did to appearance what capital did to social relations by reducing everything to the equality of objects—art became valueless and free. Although published before the Internet, Berger's notions of reproduction are even more important in the computer age that allows for extremely easy and exact mechanical reproduction. In a comment that has important ramifications for educational media, Berger suggests that if images are used in a new way, they might contain a new form of power. This use of images in different contexts is an important principle, one discussed later in the section on Surrealism.

The scholarly literature on media theory reveals an active debate over media types and their impact on civilization. In addition, this literature is preoccupied with a very specific emphasis on the effects and mechanisms of various media. The text's dominance in education is clear, but the influence of a transition on learners and society to other media is more difficult to understand. Scholars repeatedly explore how the use of specific media in various ways both reflects and reinforces power structures in society. This issue is important to consider in discussing the design of computer education environments. How are dominant social structures reinforced or challenged in the design of computer software? Many have claimed that the

Internet will lead to a democratization of learning through the easy access to information and the teacher's repositioning to a less authoritative role. Nevertheless, one needs to look more closely at how users are positioned when utilizing educational software and understand the ramifications of incorporating borrowed media, which come with their own sets of codes and social norms.

STILL PHOTOGRAPHY

Understanding the use of still photography is critically important to educational technology and the creation of computer-based educational environments. Still photographs have been used in textbooks, as instructional in-class aids, and now in computer educational environments. Because of the technical limitations of digitized video and large file size at this time, still photographs appear more frequently than moving images in computer environments. Clearly, computer software borrows and benefits from the properties of the photographic medium. What are these properties? How do they impact educational uses?

While still and moving photographs (film) have some differences, one strong similarity is the persuasive nature of the media. In the 2001 survey, respondents indicated that photographs possess some power to convince students of truths (see Figure 26). In response to the statement that "photographs are evidence of the truthfulness of argued points in the subject matter," students either strongly agreed or agreed at a rate of 68.2%.

The critical literature on still photography in many ways parallels that of cinema because of a persistent interest in commenting on the reality it mechanically records. The first still photographs taken were of the real world—family and friends often related to the photographer. According to Williams (1972), once the camera proved itself as a recording device, it inevitably came to be regarded as a tool for

Figure 26: Photographs Evidence of Truth (Questions 1 & 15)

Delivery format * Photographs are evidence of the truthfulness of argued points in the subject matter. Crosstabulation

				Photographs are evidence of the truthfulness of argued points in the subject matter.				Total
				strongly agree	agree	disagree	strongly disagree	
Delivery format	computer-based	Count		3	19	14	2	38
		% within Delivery format		7.9%	50.0%	36.8%	5.3%	100.0%
	videotape	Count		1	34	11	1	47
		% within Delivery format		2.1%	72.3%	23.4%	2.1%	100.0%
	correspondence	Count			2			2
		% within Delivery format			100.0%			100.0%
	other	Count		1				1
		% within Delivery format		100.0%				100.0%
Total		Count		5	55	25	3	88
		% within Delivery format		5.7%	62.5%	28.4%	3.4%	100.0%

changing things to the way they should be and emerged as an instrument of social change. The realization that the camera could comment on reality while recording it led to the second stage in its development, one concerned largely with making the public aware of social problems. Jacob Riis' influential 1890 book *How the Other Half Lives* spurred political reform when it showed the extreme poverty evident in New York tenements. Later, Lewis Hine's child labor photographs led to similar changes. Commissioned by the Department of Agriculture during the Depression, one of the most famous groups of American photographs ever taken were 270,000 photographs documenting the state of farmers. Many were taken by photographers who would become extremely influential, such as Dorothea Lange and Walker Evans, and embody how during this time period photography emerged as an instrument of social change.

Figure 27: Documentary Photography—Old Age (Library of Congress, Prints and Photographs Division [LC-USF34-009681-C])

Note how in Figure 27 Lange's photograph manages to seem very realistic, while at the same time it is expressive, with the man's hat dipped to the camera giving an impression of resignation and defeat. While many of her more famous photographs were close-ups, Lange in fact generally used wide and long shots to capture subjects in their environment and allow the viewer to direct attention freely within the frame. We will see later that this style was adopted by the neo-realist filmmakers after World War II. This tradition of photography with a social agenda continued for many years until America turned more inward with work such as Robert Frank's in *The Americans* (Williams, 1972).

In her book *On Photography* (1977), Susan Sontag offers one of the most interesting perspectives on still photography, arguing that photography gives viewers the sense that they hold the whole world in their heads—which leads to an existential shift in the viewer. Instead of just recording reality, photographs have become the norm for the way things appear to us, thereby changing the very idea of reality and of realism. To take a photograph is to participate in another person's mortality, vulnerability, and mutability, according to Sontag. Photography leads to an acquisitive relation to the world that nourishes aesthetic awareness and promotes emotional detachment. This notion of emotional detachment is important to understand in relationship to the politics of educational uses of photography. Photographs democratize experience to some degree by making images available to all. However, photographs also reduce and limit experience in some ways to the status of a souvenir. For Sontag, photography has a basic conservative bias—to take a picture is to have an interest in things as they are, in the status quo remaining unchanged.

Furthermore, Sontag argues that to photograph is to confer importance on a subject. It means putting oneself into a certain relation to the world that feels like knowledge and power. Certainly, this notion of the appearance of knowledge through viewing photographs has relevance in understanding the use of media in education. According to Sontag, photographed images do not seem to be statements about the world so much as pieces of it, miniatures of reality that anyone can make or acquire. Consequently, she sees the knowledge gained through still photographs as always being sentimental, a semblance of knowledge. Finally, the camera's ability to transform reality into something beautiful derives from its relative weakness as a means of conveying truth. What one learns from Sontag is that the camera is an inherently unreliable educational medium, and consequently photographs and films need to be very carefully used, and the political and social impact understood.

The reality effect of photography is crucial to understanding the use of this medium in computer-based educational environments. Much like Sontag, Rabinowitz

(1994) claims that the invention of the photograph coincides with the rise of commodity culture and was driven by a desire to own what was photographed. In terms of educational photographs, Whatley (1990) claims color photos employed in textbooks emphasize the realness of images. Generally, photographs used in textbooks starting in the 1950s had an advertising look because they took advantage of the perception of an audience accustomed to the advertising genre for coding. Often photographs in textbooks manipulate student viewers by using pictures of models posing as people that the students might wish to become. Thus photography used in educational material claims to be both evidence of truth and a commodity.

PHENOMENOLOGY OF FILM

Susan Sontag's existential notions about still photography fit well with a long tradition in the critical literature of cinema on the phenomenology of film. Since its inception, one of film theory's central concerns has been to try to comprehend the nature of representation in this unique medium. As we saw in critical commentary of those who have written about still photography, the fact that film has a physical relationship to the object it is representing makes it special as a medium. Andre Bazin, the famous French film critic, talks about this quality of cinema as a "death mask," a medium that has the ability to make a remarkably life-like impression of reality (Wollen, 1970). This characteristic of both still and moving pictures to record reality has important implications for education because, as we saw from the earlier data (Figure 25), photographic evidence persuades students of the truthfulness of particular arguments.

Kaufman and Goldstein (1976) claim that film is timeless, or tenseless in literary terms, and appears variously as thought, memory, or dream. Consequently, it is up to the audience to construct meaning out of the temporal worlds on the screen and make sense of them. For Kaufman and Goldstein, the indeterminacy of space is a key to understanding cinema's use of time because even the most realistic film style transforms reality into a world with little connection to the one we experience first hand. They argue that the film image has three basic qualities: representation, assertion, and symbol. First, filmmakers try to make images look like the objects they want to represent. Second, filmmakers assert a meaning through the use of symbols representing the real world. In other words, the objective seeming reality of the photograph is given meaning by a symbolic reference. The symbolic aspect of the image in film leads to unexpected or imperceptible meaning transmitted within a context controlled by the filmmaker. Finally, according to Kaufman and Goldstein, the film star is the primary representation made in Hollywood film. Films represent types of people who are often characterized in specific ways to encourage the audience's identification and manipulation. One important question to ask is: How might faculty members or experts in given fields serve as stars in educational media?

Clearly the power of still and moving film images—and now digital images on the computer—is in these phenomenological characteristics that convince the viewer of the reality of what he or she sees represented. Unlike written text or in-person speech, photographs appear to the student viewer as having more truthfulness and are less likely to seem to be manipulations or expressions of point of view.

I AM LYING: DOCUMENTARY FILM AND TRUTH

Having looked at the basic characteristics of still and moving images and the reproduction of reality, we now consider films purporting to document reality. On December 28, 1895, Auguste and Louis Lumiere showed their device to the Paris public for the first time in the Salon Indien of the Grand Café. A train, seeming to come right out of the screen, reportedly made the audience flee into the streets in fear. Clearly, the audience was convinced that the train projection was real. According to Barsam (1992), this incident marked the end of one kind of seeing and the beginning of another: seeing through cinema as a new medium.

Much as we saw in the early history of American still photography, the hope of non-fiction film is that it will change mankind's perception of the world. Thus, early on, documentary film often had a political agenda. Although photographic images and documentary films of various sorts have been employed in education for many years, there still is not an adequate understanding of the theoretical implications of their use. Now with computer-enabled learning, the importance of understanding the use of still and moving images in education must be better understood. Issues such as point of view, implications of editing and image association, passive versus active positioning, and the psychology of viewing images are central to this discussion. Much of the work done in the last few years in narrative and documentary film criticism is relevant and should be incorporated in the research of the educational application of still and moving images.

John Nichols (1994), a leading documentary film scholar, classifies films in a reality continuum as follows: Hollywood fiction showing an absence of reality; documentaries from the 1930s directly addressing reality in often didactic terms; documentaries from the 1960s that observe and reject commentary, but lack a sense of history and context; documentaries from the 60s and 70s using interviews, sometimes having too much faith in individual witnesses; documentaries from the 1980s questioning documentary form, often overly abstract; and documentaries from the 1980s and 1990s stressing subjectivity, sometimes to excess. In looking at the history of the documentary form, Nichols argues that traditionally documentaries suggested fact, but recently they have emphasized uncertainty and subjectivity. Importantly, he questions what position viewers take when they encounter stories about history. Once selection and arrangement occur in documentaries, issues of objectivity, authenticity, power and control arise.

Nichols finds it remarkable that documentaries so easily use narrative conventions without awareness, introducing characters and situations, showing a problem or conflict, and offering a resolution. Traditional film techniques such as establishing long-shots, close-ups as an intensifier to heighten reaction, and continuity editing are all regularly used in documentaries. Furthermore, asynchronous sound is employed in documentaries working to emphasize a narrative or storytelling format. According to Nichols, televised network news uses melodramatic codes, telling news stories as conflicts between good and evil. One can see that although documentary films present themselves as truthful evidence, they in fact often use standard fiction film devices. Additionally, Nichols sees a problem with much of the reality TV that became popular in the 1990s in that it refuses to allow a sense of historical context or consciousness. Finally he asks the key question about representation of people in documentaries: Who has the right to represent others and their points of view?

Nichols describes four general forms of documentary film: expository, observational, interactive, and reflexive. The reflexive film challenges spectator assumption by making plain its representational functions or by upsetting spectator assumptions through its political content. Plantinga (1997) argues that there is a difference between formal, open, and poetic voices of the nonfiction film—open voice is more reticent in the impartation of presumed knowledge and probably closest to educational purposes. According to Plantinga, when a film is recognized by the viewer as non-fiction, it encourages a special type of spectator activity—the filmmaker asserts that he or she presents something that occurred in the actual world. Non-fiction films are not imitations or re-presentations, but constructed representations. The unique value of documentary photography is the capacity to provide forceful information. Documentaries have a richness as a medium that often presents more that just the intention of the filmmaker. This is their particular value for education, as well as a challenge, because the meaning presented is not as easily controlled as in text. Nevertheless, the use of computers has changed how non-fiction films are regarded by the viewer in terms of objectivity. Digital imaging technology allows for manipulation that calls into question the truthfulness of photography. Are viewers more aware of manipulated photographs than in the past?

Respondents to the 2001 survey indicated that they were to a large degree aware of the point of view of the course author (see Figure 28). In response to the statement, "I am very much aware of the point of view of the author(s) in representations made with various course media," 84.6% of the students claimed they either agreed or strongly agreed.

Figure 28: Aware of Author Point of View (Questions 1 & 19)

Delivery format * I am very much aware of the point-of-view of the author(s) in representations made with various course media. Crosstabulation

			I am very much aware of the point-of-view of the author(s) in representations made with various course media.				Total
			strongly agree	agree	disagree	strongly disagree	
Delivery format	computer-based	Count	6	37	1		44
		% within Delivery format	13.6%	84.1%	2.3%		100.0%
	videotape	Count	1	33	12	1	47
		% within Delivery format	2.1%	70.2%	25.5%	2.1%	100.0%
	correspondence	Count	1	3	1		5
		% within Delivery format	20.0%	60.0%	20.0%		100.0%
	other	Count	1				1
		% within Delivery format	100.0%				100.0%
Total		Count	9	73	14	1	97
		% within Delivery format	9.3%	75.3%	14.4%	1.0%	100.0%

Clearly, students believe that they are conscious of author bias as an issue in media. Given the power of the medium, one wonders if this confidence in their own awareness of bias expressed by the students is justified. Are they really aware of the author's point of view in the course material? The theoretical literature on documentary film is illuminating here because issues of author perspective and the manipulation of recorded reality are at the heart of such discussions.

Rabinowitz (1994) claims that documentaries construct a version of the truth and a way of seeing, but that the narrative sequence often requires the viewer to understand the truth represented in conventional ways. She argues that the documentary is very invested in what she calls "narrative forms of difference," which focus on the question, "Who is looking?" She points out that the dictionary defines "history" as a narrative, while documentaries are concerned with "evidence." Consequently, she sees documentaries as a means of instructing through the presentation of evidence in various ways.

Documentary film theory is directly relevant to an understanding of computer-based educational environments as a medium. As the evolution of the documentary film has led to an increasing consciousness of subjectivity and the implicit ideology transmitted in non-fiction works, students report an awareness of the point of view of media authors in courses. One wonders, however, if this is the case. Although there is some evidence to indicate that the widespread use of digital images has led to an awareness of ease in altering photographic evidence, some scholars have emphasized building an increased viewer consciousness of representation of point of view through the style of documentary media. This approach makes a great deal of sense for use of media in educational environments where the development of critical thinking skills and awareness of author perspective is crucial.

ALTERNATIVE STRATEGIES

Building on what we have learned about non-fiction and fiction theory that is relevant to understanding educational applications of computers, it is fruitful to look at major alternative strategies used in media. In opposition to traditional dramatic structures which some describe as illusionist, there are two theoretical traditions outside of Hollywood useful to understand: the theories of Bertolt Brecht, and the Surrealist Movement.

Bertolt Brecht, the famous German Expressionist playwright, espoused a dramatic theory that emphasized preventing the audience's tendency to identify with characters on the stage. Coming from a leftist political perspective, Brecht argued that if the audience believes the illusion of the reality of fictional works, they are in turn prevented from thinking critically and analyzing the subject matter. He therefore employed various techniques—including breaks in the dramatic action as well as extremely exaggerated styles of acting and set design—to make the audience continually aware that they were watching a fictional play. This technique of distancing the audience was a technique later used by the French New Wave filmmakers—especially by Jean-Luc Godard, beginning with his film *Breathless*. Very much aware of Brecht's theories of distanciation, for both political and aesthetic reasons, Godard wanted to thwart the fantasy experience of the audience. Thus with jump cuts (gaps in the editing of scenes), exaggerated camera movement, and self-reflexive action, Godard intentionally denied the audience the illusionary experience of the Hollywood-style film. For Godard, this dramatic theory of film makes the subject matter of almost all of his films essentially political and self-reflexive; the themes of many of his movies are directly about the *form* of film. Would it be useful to make educational software that is about the form and structure of the learning experience, or at least encourage an awareness of the form in which knowledge is transmitted?

The Surrealist Movement combined Freudian theories with a radical political perspective. The most notable Surrealist filmmaker in this tradition, Luis Bunuel, is interesting to consider here because of his deeply philosophical approach. His film style incorporated fictional dreams and the language of dreams in the narrative structure. As a Surrealist, Bunuel played with context by taking objects from one realm and placing them in another to create new meaning. His aim was essentially artistic: he wanted to give the audience a different perspective and to disturb them with strong images—from an eyeball slit with a razor, to a beautiful woman with only one leg.

Almost all of the premises for his films were based on profoundly disturbing subjects: from an upper-class woman who feels the need for abuse as a prostitute (*Belle De Jour*), to a group of party-goers overcome with a collective fear of leaving the house (*The Exterminating Angel*). Although it would seem strange to many, the role of distancing and challenging the audience needs to be considered in education more often. We will see in Chapter Ten how Bunuel's techniques are used to take advantage of the similarities between watching moving images and dreaming. In particular, the Surrealist preoccupation with violently ripping objects from contexts is one that gets at the heart of the meaning-making activity. In education, students need to understand how context affects meaning, and the use of Surrealist strategies in educational environments is one way this can be accomplished.

Both of these traditions employ different strategies for dealing with the audience's inclination to perceive a film as reality. They both have artistic, rather than commercial, intentions. With Godard, his strategy is to work to prevent the narrative process altogether; with Bunuel, on the other hand, he utilizes the narrative structure and impression of reality in the film language in order to confront and disturb the audience. Even for those interested in constructing mainstream educational software, the techniques employed by Bunuel and Godard can be useful. Look at the way that various Hollywood filmmakers have utilized a self-reflexive style for both dramatic and comic effect in *FX* and *The Stunt Man*, as well as on television with *Saturday Night Live* and *The Garry Shandling Show*. Many Hollywood filmmakers have used Surrealist techniques very successfully to create a more complex style in films such as *All That Jazz* and *Apocalypse Now*, as well as on television in shows such as *Ally McBeal* and *Twin Peaks*.

The application of these two alternative strategies in educational software is more challenging, but potentially more fruitful. First, particularly for adult learners, a constant awareness of the structure of meaning systems represented in education software would prove very useful, both to increase learning and to develop critical thinking skills. The second alternative approach is an artistic one that would shift education software from the transfer of knowledge to the inciting to think more critically. As one can see, both of these methods involve the use of higher order thinking and are well suited to computer learning environments.

KNOWLEDGE MEDIA AND MIXING MEDIA

Moving from film to computer theory, it is useful to look at broader notions of using combined media for educational and non-educational purposes. The research literature shows that in addition to adopting alternative theoretical approaches, using media in combinations for various effects is an additional strategy used in computer-based learning environments. Reilly (1999) found in a study that users clearly prefer a mix of media including sound and text. How does a mix of media in educational programs affect the successful pursuit of learning objectives?

The British Open University has been a leader in looking at this issue of combining media for educational purposes, coining the label "knowledge media" to describe its approach (knowledge media was discussed in Chapter Three in reference to the group method). The phrase "knowledge media" was first used by Mark Stefik to describe the profound impact of using artificial intelligence technology with the Internet. Eisenstadt and Vincent (2000) use this phrase more broadly to encompass the process of generating, understanding, and sharing knowledge using different media. They argue that knowledge media is a dynamic process involving storing, sharing, accessing, and creating knowledge. What this means for pedagogical practice is a move away from the presentation of knowledge and towards shared knowledge processing experiences. The knowledge media approach is to understand the learners' needs, and then using research, develop solutions based on a mixture of media.

The research literature at this time does not answer the question of whether text or voice-over sound files are most effective in learning situations. Using text and speech combined has the advantage of realism, dual modality, and learner control. However, using both text and repeated visual text is redundant. In addition, audio and speech are transient unless the user has a playback control. Is this redundancy useful? Furthermore, Reilly (1999) found in a study that men tend to be less happy with recorded speech in learning programs than women. He recommends that taped audio be broken into sections and be easily controllable. Research focusing on specific effects of combining media for educational purposes such as these is growing, but clearly much more still needs to be done.

Sumner and Taylor (2000) point out that managing the media mix is of key importance in developing distance learning. As film and video are media that combine sound and moving images, the literature on these is especially relevant to creating educational software. In particular, the recent criticism that focuses on the nature of signs and meaning in film and video is especially relevant to understanding the knowledge media used in educational environments.

RECENT FILM THEORY: SEMIOLOGY

Since the late sixties, there has been considerable interest in the semiology of cinema, viewing film within the general science of signs. A science of signs has great relevance to computer-based educational software where multiple languages and symbolic systems are combined. Roland Bathes, Christian Metz, Pier Paolo Pasolini and Umberto Eco were leaders of this movement in the 1960s and 1970s and served as main influences for later critics working within this tradition. Ferdinand de Sassure, in his *Course in General Linguistics* (1959), described the science of semiology and provided the philosophical background for much of this original criticism. Influenced by Emile Durkheim, Saussure viewed signs and language from a social perspective. A second important influence was Charles Sanders Peirce, who presented a taxonomy of the sign with the following categories: icon, index, and symbol. According to Peter Wollen (1970), Peirce meant his classifications to be understood as non-exclusive. In fact, Wollen argues that the richness of cinema comes from successfully combining the three semiological dimensions of the sign together in cinema. This perspective is significant in light of the discussion in the previous section on knowledge media and mixing media. The contribution of semiology as a field led to an understanding that all meaning conveyed by signs is precluded by the structure of language. Consequently, when students encounter symbols in computer-based programs, they are always understanding them through the filter of the symbolic structure.

Jacques Derrida, in *Speech and Phenomena* (1973), presents some important semiological notions that are relevant to media. Derrida's book is a critique of Husserl's theories on phenomenology and language. Husserl's theory allows for the possibility of pure perception of empirical reality. Derrida rejects the possibility of pure perception, because it is always "perception as"; consciousness as language always intervenes in the act of perception. For Saussure, the difference between signifiers (symbols) is what constitutes them as individual signs. Language itself is founded on the distinctions between the signs: what defines a word is its difference from other words in the system. Derrida coins the term "difference" to describe the causal agent in the creation of meaning. "Difference could be said to designate the productive and primordial constituting causality, the process of scission and division whose differings and differences would be the constituted products or effects" (Derrida, 1973, p. 137).

Difference as the basis of meaning and signs argues against the possibility of pure perception—because language is founded on this principle of difference, it can never sense the real world.

> It is at the price of this war of language against itself that the sense and question of its origin will be thinkable....A polemic for the possibility of

sense and world, it takes place in this "difference," which, we have seen, cannot reside in the world but only in language (Derrida, 1973, p. 14).

The elements of the empirical world are not structured through their inherent dissimilarity, and therefore cannot be authentically perceived by language.

Traditional concepts of the relationship between language and empirical reality reveal an idealism that connects form with language, and substance with physical being. In this manner, Derrida exposes the Platonic base of Husserl's language theory that equates word with body and meaning with soul. "What governs here is the absolute difference between body and soul" (Derrida, 1973, p. 81). Derrida's criticism of Husserl points out that the concept of pure perception, and the possibility of perception's connection to signification, is built upon a concept of constant transition in time, from primordial impression to memory and expectation, thereby admitting non-presence into the relationship between empirical reality and the signifier. According to Derrida, once Husserl admits time into the notion of signification, he has shifted meaning into human consciousness, away from empirical reality: "The fact that non-presence and otherness are internal presence strikes at the very root of the argument for the uselessness of signs in the self-relation" (Derrida, 1973, p. 72).

The importance of Derrida's analysis to the discussion of film as a language is clear: every film image is constituted in time, by the viewer's expectations. Although the physical/chemical relationship between the filmed thing and the photographic process might be objective, the human viewer cannot view the image that is created without preconceptions and the memory of the images that preceded it. It is precisely "in time" that the viewer loses any chance of a direct contact with the empirical world through the photographic image.

For Derrida each speech act, which is necessarily located in time, evokes the control of the demands of consciousness as it listens to itself speak. In this way the speaker constantly corrects himself in the act of speaking. Speaking is always "speaking to oneself." The speaker always listens to his own speech, thereby keeping the language under the control of interior, meaning-giving intention. Speech never really encounters the phenomenological world, because it is constantly controlled, and listened to, by the speaker.

> When I speak, it belongs to the phenomenological essence of this operation that I hear myself at the same time that I speak. The signifier, animated by my breath and by the meaning-intention…is in absolute proximity to me.…It does not risk death in the body of a signifier that is given over to the world and the visibility of space (Derrida, 1973, pp. 77-78).

In this way the image-content in film, which can be likened to the speech act, is always construed by the filmmaker, who is listening to himself speak. He is conscious of structuring the discourse around the requirements of the language. As the filmmaker hears himself speak in the filmed utterance, the viewer in turn also repeats in himself the hearing-of-oneself-speaking precisely as he sees it in the film.

> To speak to someone is doubtless to hear oneself speak, to be heard by oneself; but, at the same time, if one is heard by another, to speak is to make him repeat immediately in himself the hearing-oneself-speak in the very form in which I effectuated it (Derrida, 1973, p. 78).

Although this might sound like philosophical double-talk, the point is important when considering how one learns through language and images. The empirical reality recorded objectively by the photographic process becomes, through the requirements of the language structure of the human mind, removed from physical reality because the listener becomes a speaker in the act of hearing. He always restricts the articulation to the sphere of non-empirical language. It is precisely the intervention of the listening consciousness that has destroyed the possibility of language being informed by the empirical world. Every moment of listening is always already committed to understanding each element of the preexistent language system.

The danger of a meaning structure removed from the phenomenological world is made clear by Martin Heidegger (1962). He describes a quality of language termed "theyness" in which individual meaning becomes suppressed by the dictates of the collective structure. Language never expresses the individual uniqueness, only the general rules of language. Heidegger declared that we must search for the "authentic" word that will bring together speech and authentic being. Cinema isn't this new authentic language that some hoped it would be. Each image of a film is always already constructed in the language pattern of the viewer, and this has deep implications for education.

What does this extensive and dense philosophical semiological literature mean for those creating computer-based educational environments? First, one needs to be conscious of the nature of signs. In particular, semiology points out the approximation of truth and meaning that all symbolic systems employ. New media often employ a collection of symbolic systems including text, photographs, video and film, drawings, graphic elements, and sound. Especially with photographs and film, the illusion of direct truthfulness is strong and represents a manipulation of the learner/viewer. Saussures' notion of meaning as coming from the differences among symbols is important in developing teaching strategies because it emphasizes the way meaning is negotiated through planned distinctions between individual symbols and concepts. The essential conservative nature of symbolic systems is vital to understand, as well as the notion that these systems are social constructions. As such, symbolic systems used in educational software are by definition deeply rooted in socially accepted meaning and conservative notions of truth.

COGNITIVE FILM THEORY

In recent years, film theory has seen the emergence of a cognitive theory of narrative. This film theory is relevant to computer-based educational software because it focuses on the use of narrative and how it might be employed more effectively to increase student learning. David Bordwell, Edward Branigan and Noel Carroll analyze cinematic comprehension in terms of an active viewer. They claim that film viewers use the same basic psychological processes in viewing a film as in understanding the world around them.

Branigan (1984) emphasizes the cognitive role of the viewer, which is often minimized in psychoanalytic models of film. Narration is a logical relationship posed by the text as a condition of its intelligibility. For Branigan, the viewer actively constructs the space of a film. In comprehending or interpreting the screened images, viewers actually construct the film. Branigan argues that psychoanalytic film theory has ignored the importance of the cognitive skills of the viewer. He defines narration as a linguistic and logical relationship posed by the text as a condition of its intelligibility. Branigan concludes that the viewer constructs the space of the film. In a similar fashion, students using media construct their own personal film.

For Bordwell (1989), meanings are not found but made. As a set of psychological processes, comprehension is an activity of meaning making. Cognitive psychology, he claims, best reveals the nature of those meaning-making processes. In holding that the film viewer actively realizes the film's meaning, Bordwell claims that the film, at least in its semantic content, is incomplete, only partially created. The film image is necessarily incomplete so that it can be fleshed out by the active participation of the viewer. By insisting on the semantic incompleteness of the film, he seeks to emphasize the necessary role of the active viewer. Bordwell argues that the cognitive revolution offers a useful framework for analyzing film. He uses what he terms a constructivist theory of interpretation for film whereby he believes viewers construct meaning from symbolic cues when viewing a film. This principle of incompleteness can be used in education to involve the learner.

According to Bordwell (1997), cinema is a communication medium first, an art only secondarily. Bordwell quotes Noel Burch in naming the illusionist cinema of Hollywood the "Institutional Mode of Representation" (IMR). Building on the phenomenological theory reviewed earlier in this chapter, he suggests that the film medium came about because of a desire to recreate a perfect illusion of the perceptional world. Thus what began as a machine for reproducing perceptual reality became a vehicle of fantasy. Of all media, only film could completely and unobtrusively fulfill the bourgeois dream of replication. A film's style then becomes a strategy for creating an illusion of reality. Bordwell defines style as a film's systematic and significant use of techniques of the medium including mise en scène (staging, lighting, performance, and setting), framing, focus, control of color values, cinematography, editing, and sound. He sees style as being linked to the psychology of each cinematic period. A history of film then is the development and evolution of fantasy forms and stylistic approaches to this creation of fantasy worlds. How might these fantasy worlds be used to increase student learning in educational software?

For Carroll (1988), a viewer internalizes the interests of the characters and assesses the series of options possible. These subconscious expectations are formed as questions. Building on the Russian director Pudovkin's theories, he suggests that the relationships, between scenes in a film is one of questions followed by answers (later, this notion is explored further). Carroll believes that the viewer internalizes the structure of the film, including various plot options that may be presented. Can this internalization of question-and-answer narrative structure be used to good effect in educational software?

Cognitive film criticism fits with a general movement in the analysis of cinema as a medium that focuses more on the viewer than the filmmaker. The psychological processes of the viewer in watching a film are not unlike those of a student using a educational software product. Taken collectively, these theorists argue that through the incompleteness of the film medium, viewers construct their own internal movies, often using question-and-answer story structures to create fantasy worlds that are convincingly real. In an educational environment, if one could effectively control the internal movie construction of students, learning might be greatly increased.

CONCLUSION

Many films have a "reality effect"—a mask of reality. Films based on true stories (*In Cold Blood*, *Shindler's List*), historical pictures (*Glory*, *Pearl Harbor*), and issue films (*Silkwood*) often constitute this type. The more we are convinced of the "real" nature of the film, the greater the effect it has on us. This principle is even more important in educational software than in fiction film. It is interesting in this context to discuss the Italian neo-realists such as Rossellini and DeSica who developed a "reality style." Their post-Word War II films are very powerful and had a great deal of influence on later filmmakers such as Antonioni, Fellini, and Truffaut. Of course, the neo-realist films were hardly documentaries—

they used scripts, trained actors, and sets. But they developed a "style" of realism by employing techniques such as long takes and deep-focus. In addition, they chose simple, often lower-class, stories to tell (such as in *The Bicycle Thief*, where a man searches for his stolen bicycle that he needs to keep his new job). However, the important point to remember about the neo-realists is that films such as *The Bicycle Thief* and *Open City* were powerful to a large extent because they were able to convince the viewers of their authenticity, of their "realness."

The nature of the film process lends itself to an assumption of realness. Most viewers are manipulated by this reality effect. Although some filmmakers and still photographers have tried to honor the reality effect by approaching their art in alternative ways, they still end by creating yet another style, one that may in fact manipulate the audience more by increasing the lifelikeness. In designing educational software, one needs to be aware of the difficulty of dealing with the learner's tendency to both be convinced of the realness of what is portrayed and to connect images with personal fantasy structures. These two tendencies are discussed in more detail in the next two chapters.

In this chapter I investigated the notion of computer environments as media. I then turned to an examination of various relevant research literature on media, photography, and film. The literature on film theory in particular was reviewed because it has a strong connection to computer environments. Now that the background of computer as medium has been sketched out, we turn in the next chapter to the specific issue of the use of dramatic structure, genre and the ordering of images through editing.

Chapter X

Dramatic Structure, Genre, and Editing

We saw in the previous chapter how understanding that the computer is a medium, not just a tool, reveals film theory's relevance to the discussion of computer interface design. Since the computer as a medium is in an early stage of development, some interface designers have looked towards film for structural models and specific techniques. Computer applications such as Macromedia's Director, the dominant multimedia authoring program, explicitly employ the film metaphor, both in its conceptual design and its use of terminology such as scene, script, score, and stage. Additionally, computer applications increasingly draw on digitized film and some simple elements of its grammar, such as transitions, framing, camera movement and camera angles. A more mature medium, film may supply some of the best approaches to computer interface design.

Lydia Plowman (1994) notes similarities between early film and interactive multimedia in establishment of narrative conventions such as intertitles and narration. According to Plowman, the audience needs to make a cognitive shift in order to adapt to new media. She quotes Noel Burch (1981) in describing the transition from early to mature forms of film involving a linearization of the narrative. Early film emphasized spectacle and the documentation of unrelated events. Events and individual shots were not woven into a coherent narrative until D.W. Griffith and others led the way to the development of montage and a cinematic narrative language. Plowman argues that this same process of making new media more coherent needs to occur to increase the educational effectiveness of computer-based programs.

Furthermore, computer instructional designers do not have ready access to an established narrative language and consequently need to be more explicit in their structure. The audience's knowledge of film conventions allows the authors to feel confident that their narrative can be quickly and simply understood. Consequently, Plowman argues that instructional designers need to spend time developing narrative conventions and make narrative elements clear to the learners. In this chapter we look at narrative or dramatic structure and image order (editing) in film and other media and point out key issues in the use of sequence strategies in the design of computer-based educational environments.

USE OF DRAMATIC STRUCTURE

Brenda Laurel and Janet Murray offer some of the most intriguing theories about the use of dramatic structure in computer software. In *Computers as Theatre* (1993), Brenda Laurel examines how computer interfaces might be best constructed as dramas in the tradition of Aristotle's theories as represented in the *Poetics*. As is my position in this book, Laurel sees the computer as a medium, not a tool. Consequently, the computer environment needs to be analyzed in regard to specific principles as a new medium, just as occurred in the development of photography and film. Laurel defines theatre as representing whole actions by multiple agents; she draws a direct analogy to computer interface design, with direct manipulation or engagement a key aspect of interactivity in the computer. For Laurel, there are two primary advantages to thinking about computers as theater: significant overlap of actions through the use of agents, and the familiar and evocative nature of theater in the interface. Here Laurel's notion of the "familiar" intersects with notions from cultural psychology of an established canon all members refer to when interpreting stories. Laurel distinguishes "drama" from "narrative" by stating that drama is more active, intense, and has greater unity of action in the Aristotelian sense. She concludes that interface design should focus on action.

Janet Murray (1997) argues that stories define how we think, play, and understand our lives, and sees computers as having a profound affect on the collective cultural stories created in the late 20th century. Murray asks a key question about the use of stories in interactive computer environments: How can we enter the fictional world without disrupting it? She points out that computer-based narratives seem to show the tendency to emphasize the border of real/unreal and test the dramatic illusion. Murray proposes a notion of interactive computer narrative as labyrinth, goal driven enough to guide navigation, but open-ended enough to allow free exploration. She argues that the navigational space lends itself to journey stories, and that the computer has transformative power that leads people to assume roles. Digital stories refuse closure and tease the viewer with incomplete endings. But, how do you have catharsis in a medium that refuses closure? Murray asks if it is possible to have both freedom and closure? Can there be destiny in a plotline with variables? She argues that formulas in storytelling are well suited to digital storytelling, but the ending of a story is key—it must have causality and a kind of logical determinism. Finally, Murray sees the most ambitious promise of the new narrative medium as its potential for telling stories about whole systems with vivid and encyclopedic detail.

Both Laurel and Murray point to the uniqueness of computers as a storytelling medium, with Laurel pointing to the active and familiar nature of stories as a key advantage for use as a computer interface. Murray broadens the discussion by focusing on interactivity in controlling the direction of the plotline, noting that how resolutions are handled is still a fundamental issue.

NARRATIVE CONVENTIONS: PREMISE AND THEME

Film and television, as with many other popular art forms, possess various kinds of conventions. Conventions in narrative structure are extremely common and provide the backbone for most Hollywood-style films. In Hollywood, a story is usually bought or sold on the basis of its "high concept" or premise. High concept is the story summarized into a one-sentence sales pitch, such as *Rocky*: A third-rate boxer gets one shot at the championship; or *Close Encounters of the Third Kind*: A workman is led by intuition to a UFO landing site that has been concealed from the public. Conscious attention to this notion of high-concept films has emerged with cartoon-style films such as *Star Wars*, *Jaws*, and *Raiders of the Lost Ark*, as well as films directly based on comic books, such as *Superman*, *Batman*, and *Dick Tracy*. Reportedly, such influential filmmakers as Steven Speilberg and George Lucas are said to only be interested in stories that can be summarized in one sentence. But this notion isn't really anything new for Hollywood. Hollywood-style scripts have always emphasized simple stories told in a direct, obvious fashion, consisting mostly of external conflict. Aristotle and Lejos Egri are two of the most influential figures in expressing the rules of dramatic structure known in Western culture. Their theories are fruitfully examined in relationship to instructional design because they form the basis for learner expectations when using narratives in educational software. When software designers borrow from film and television, they utilize Aristotle's and Egri's dramatic language.

The Art of Dramatic Writing (1946) by Egri is a book with which most creative writing students are familiar. Along with Aristotle, Egri is constantly cited as the expert on the three-act structure. His point of view is interesting to analyze because in many ways he represents the dominant theory of the premise, or of "high concept" as it is called in Hollywood today. Egri tells us that there is a three-part structure to the premise sentence: character, action, and ending. For computer environments, this structure corresponds to agent, interaction, and change or learning.

Figure 29: Three-Part Dramatic Structure

Egri	Character	Action	Ending
Computer	Agent	Interaction	Learning (change)

Clearly this formula embodies the usual linear, egocentric, Western notion of story structure. Someone (you the audience) takes an action (complication, conflict) that leads towards a goal (wish, ambition). According to Egri, all Hollywood-style films have a simple formula structure, this plain enunciation of their themes. These stories are always about someone: a main character, a star. The premise sentence must include a primary action. This action describes the main movement of the story, the primary direction. The main character learns, struggles, fights, works, travels, runs, and discovers. The verb is the essence of the premise sentence. For Egri, the premise should never be contemplative—stories are about actions: "The moment you decide upon a premise, you and your characters become its slave. Each character must feel, intensely, that the action dictated by the premise is the only action possible" (Egri, 1946, p.151).

Aristotle takes a similar position: "All human happiness or misery takes the form of action; the end for which we live is a certain kind of activity, not a quality..." (Aristotle, 1941, p. 1461). In many ways, traditional Western story structure is the theater of "What happens next?"

In looking at the educational applications of this theoretical proposition, one should consider the role of theme in story structure, as this is often the focus in stories used for education. How is the theme developed in relation to the plot? What is explicitly and implicitly conveyed in a story about the theme? In a traditional literary sense, an authentic theme comprises some identifiable truth about the human condition revealed to the audience. On a psychological level in film, an authentic theme is one that taps into a desire or wish of the audience/viewer. An authentic theme for educational purposes would involve the seeking of truth, but probably without the depth of a literary endeavor. A theme described without a plot sounds trite and so general as to be meaningless. A premise has to have the specific clothing of a series of actions in order to create any kind of real meaning. Likewise, educational uses of story need this same specificity.

European-style films tend to be character-based rather than plot-oriented, enabling them to unite plot and theme more closely. A film like Truffaut's *Four Hundred Blows* succeeds at combining the theme of a childhood search for love and belonging with the action of a lower-class boy roaming the streets of Paris. Such unity of plot and theme occurs here because that film's focus is completely centered on the character rather than on the plotline. Hollywood-style films often have a gap between action and theme, so that they almost create two separate movies. In Hollywood-style films, the theme is usually carried in the subplot, or B-story. The subplot generally has its own abbreviated three-act structure and often occurs in the script as a moment of pause in the action, a skip in the relentless drive of the plotline towards the climax. When one describes the theme of the film, one often ends up describing the subplot more accurately than the main plot line. For educational software designers, a more character-based approach to the use of narrative might be more successful because theme and plot is more closely intertwined.

An example of how theme and plot is used in educational software is found in the popular *Oregon Trail* application. This software has a clear plot, the journey West, and has specific characters representing certain classes, genders, and types

Figure 30: Theme and Plot Diagram

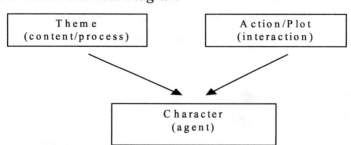

of pioneers. While the characters certainly add texture to the lesson (the 4th edition contains video vignettes of each character), they are not incorporated them into overall themes. In order to change this, the designer would need to have the plot altered by the choice of character, not just using them as resources for gaining points along the trail West.

What does this gap between theme and action in the premise mean to the audience? The further the plot line is removed from the theme, the weaker the affect on the audience. In the Hollywood-style film, the best premises are the ones that incorporate the theme into the action. One reason for the success of *Citizen Kane* is the way that the main premise of the reporter searching for the meaning of Kane's last word becomes an investigation into the theme of how power corrupts. Plot is always manipulation of the audience. In striving for high-concept, plot-oriented stories, Hollywood filmmakers often leave the themes behind. The theme is the substance of the story and of learning. Good, effective premises are those that manipulate an audience towards the experience of an authentic theme. In constructing educational software using story, it is important to be aware of theme and action. Simply creating plot and characters to navigate through content isn't enough. Designers need to mesh the theme with the action through strong characters.

GENRE AND AUDIENCE EXPECTATIONS

What expectations do users of computer-based educational programs have that color their experiences and perceptions? More broadly in other media, there are expectations formed by various genre forms. Most spectators walk into a movie theater expecting a certain type of film, as do readers when they pick up a mystery or other category fiction. According to Hedges (1991), film spectators rely on their specific knowledge of other films when encountering a new work. Commercial filmmakers are very much aware of this viewer tendency and the desire of audiences for familiar forms and try to meet viewer expectations. One principle is that the more original the film content, the more conventional elements it needs to make it accessible to the audience. This might also be true of educational software.

Branigan (1984) points out that film spectators, through past exposure to films, know how to understand new films. This ability to understand films through viewing competence needs to be understood, particularly as it relates to educational media. Branigan suggests that narrators or authors can delegate the carrying of point of view responsibility to characters or specific perspectives as a kind of framing of the story. This technique is typically employed through the use of a main character through which the story is told. Specific cinematic techniques expressing this use of a specific character to provide a point of view are described in the critical literature as "motivated" or "unmotivated" camera angles, meaning camera angles that seem to come from a character's perspective.

In looking at the use of audience expectations, the development of the American screenplay is useful to understand because it embodies the evolution of a new popular media form, one that educational media may parallel to some extent. Overall, the history of screenplay style reflects increasingly complex social influences. Silent films were part spectacle, part historical document, and fantasy. With the coming of sound, the 1930's screenplay was influenced greatly by the Depression, New York social realist writers, populism, and the development of American-style storytelling. Comedy was particularly popular as a form and helped to shape the 1940s screenplay; during the war years, scripts often had a political angle, and a new stylistic leanness. With the rise of the Cold War, the 1950s screenplay became concerned with subtext, stories expressed mood and tone more than the simple plot-line conveyed. Additionally, radio and television with episodic formulas influenced the form. The screenplay of the 1960s involved new methods and new content issues. Knowing that they had a audience trained by years of film viewing, filmmakers began to use a narrative shorthand. One can see from this short history of the American screenplay the way in which other media, the changing social-political environment, and growing maturity of the form led to the present day set of conventions.

While there are standard "rules" and conventions, the craft of filmmaking offers a filmmaker many choices. But the fact is that even a hard and fast rule such as the three-act structure has alternatives. Even if writers follow the established wisdom in creating their scripts, which these rules represent, it is crucial that they understand the devices used and the choices made. When beginning screenwriters sit down at the word processor, the first things that often comes to their minds are the old television programs and films that they've seen. Almost invariably, their stories take the shape of the storylines stored in their memory, perhaps combined with something of personal interest and meaning such as a coming-of-age story or a story about family relationships. Often, beginning writers are then exposed to a series of uncritical rules about writing that fall in nicely with those old television shows and Hollywood films— and they are on their way. The problem is that these beginning writers have no idea that in this process they have just made a whole series of choices, or more accurately, not made them. Consequently, the style and crafting of their scripts are often as derivative as the content of their stories; they are not in control of their tools and techniques. In the same way, educational software designers need to be conscious of the choices they are making and not simply adapt conventions from other media without understanding the ramifications.

Whenever a computer educational program employs video or a series of images, it automatically calls up established film conventions in the student's mind. Designers must be aware of these conventions and in the future will want to develop their own. Even in the short time the Graphic User Interface (GUI) has been in place, some unique conventions have developed, such as underlined text being hyperlinked and the rolling of the mouse over images that reveals text that describes the images. Note here that these two conventions both have to do with interactivity, and I suspect that many of the new conventions for computers will have this focus.

CHANCE AND RANDOMNESS

One illusion that is always implied in fiction film is the sense that what happens next on the screen cannot be foretold, that in some way the plot line mirrors the randomness of real life. When considering this notion in relation to the previous discussion on genre and film conventions, a paradox emerges. On one hand, through the use of formulaic plots and genre conventions, the viewer knows what to expect, and the pleasure derived from viewing is related to the degree that those expectations are met. On the other hand, the chance that something unusual or different might happen, as in real life, keeps the viewer's interest and helps in convincing the audience of a reality effect. What role might chance play in educational software?

Noel Burch (1981) describes the feeling of the viewer of an uncontrollable world just out of the film frame. The chance implied in this film world is a controlled chance. Non-narrative and experimental films often play with this sense of chance and randomness as an intentional way of examining the viewer's expectations and belief in the lifelikeness of film. The implications of the use of chance and randomness in educational software are important for two different reasons. First, the possibility of chance in conventional media is tied strongly to the reality effect leading the viewer/learner to be convinced of the truthfulness of what they see. Second, the intentional use of randomness in educational media is a strategy that might help learners make unusual and powerful associations. By intentionally programming random associations within a computer-based educational environment, designers have a powerful tool to spark thinking.

EDITING

It is hard for contemporary film and television viewers to appreciate how artificial the editing of film appeared to early audiences. We have become accustomed to mentally piecing together film shots from different camera positions—a particularly easy process for those conditioned by watching year after year of television. This same conditioning helps students understand editing conventions in computer environments. Editing, or "montage" as it is known by film critics, is the process of assembling different shots together in some sort of meaning-giving order. Although rudimentary editing strategies began earlier, D. W. Griffith is often credited with being one of the first innovators in the effective use of editing, establishing conventions still in use today. Mitry (1997) describes Griffith's strategy as structuring shots in an appropriate rhythm with the aim of telling a story and having an affect on the audience.

In film criticism literature generally, the early Soviet filmmakers are credited as an important influence on editing theory and practice. Before the 1917 Russian Revolution, the idea of montage was in the air through formalist ideas of fragmentation and recombination with artists such as Vladimir Mayakovsky, Vseyolod Meyerhold, and Yakoiv Protazanov (Barsam, 1992). Lev Kuleshov's writings were particularly influential because they paved the way for the first theories of cinema removed from theatrical bias and towards an understanding of the new medium's specific capabilities. In a famous experiment, Kuleshov used a close-up shot of a Russian actor's expressionless face as a reaction shot in three different sequences. The actor seemed to be reacting to a bowl of soup, a woman in a coffin, and a child playing with a toy bear. Although the actor appears to be reacting to each different scene, it is the exact same reaction shot. The so-called "Kuleshov effect" is the ability of film to juxtapose images through editing in a way that changes meaning. Thus a picture of a man followed by a picture of a bowl of steaming soup clearly communicates hunger. Film images allow one to recognize or relive an experience in life according to Kuleshov, and this accounts for the power of these editing effects.

EDITING STRATEGIES

The Kuleshov effect represents the general principle that nearly all editing strategies in narrative film are devised to set up a framework of expectations through a series of shots. The result is what Katz (1991) calls narrative motion. How might these expectations be manipulated for educational effect? While a number of specific editing techniques have emerged over a century of filmmaking, a look at some specific strategies with particular relevance to educational environments is illuminating.

In a general way, Mitry (1997) quotes Georges Sadoul in identifying three main editing effects: alternating of close and long shots, following a character's movement from one location to another, and alternating between episodes occurring in different locations. Note that all of these editing effects imply an audience point of view or positioning. What has become known as "seamless editing" developed in Hollywood films in the 1930s and 1940s and made these basic editing effects subtle forces of viewer control. Originally, seamless editing appeared as a kind of realistic approach by avoiding calling attention to itself and the film apparatus. We saw in the previous chapter's discussion on the phenomenology of film that this question of realism is one of the prominent issues in film theory. The physical reality of the photographic image complicates the notion of realism in film. What is the connection between editing styles and Bazin's death mask notion of film? I see two ways that realism is implied in the editing process: selectivity and point of view. First, Hedges (1991) points out that selectivity is a part of filmmaking and photography in as much as decisions are made about what to include and exclude in the frame, as well as what is left on the cutting room floor. Second, as we will see in the next chapter, editing is a primary vehicle for the representation of subjectivity and audience point of view.

The use of specific techniques in transitions and cutting on action are two clear areas where film technique and educational software connect. One obvious application of this parallel is in transitions such as the cut, fade in, fade out, dissolve, wipe, overlay, and cinematic effects such as multiple exposure, panning, zooming in, and emphasized camera angles (Baecker & Small, 1995). However, some feel that film-style transitions are overused in computers and employed without specific intentional meaning in current software (Westland, 1994). Transitions have very specific meaning in effective films, and this coding needs to carry over to educational software. Katz (1991) points out that cutting on the action is found in all types of film sequences. One of the most consistent principles used in film and video editing, cutting on action also applies to the use of music. Hollywood musicals, for example, often use the practice of cutting on the beat of the music to more closely synchronize the musical and visual effect. In computer environments, where users often control when there are transitions between "scenes," designers need to think about how to use transitions and cutting on action to increase learning.

These various techniques raise questions about their use in educational environments. Designers should consider the following: How seamless should the editing be? How should viewer expectations be set using editing to increase learning? Westland (1994) argues that the parallel editing style of film—where two sequences are intercut in order to build tension and contrast content—should be used to organize content in software. Should parallel editing be used in educational software products? How does cutting on action manipulate the audience?

EISENSTEIN

Soviet filmmaker Sergei Eisenstein is widely regarded as the first and most important major cinema theorist, particularly in his concentration on montage (Kaufman & Goldstein, 1976; Wollen, 1970). For Eisenstein, montage is the sequence of images in a film—the way it is edited to achieve specific effects. He worked to restructure film time so that it is reordered for dramatic effect. Not only can time be contracted in film, but Eisenstein showed it can be expanded—as in his famous Odessa steps sequence, for instance. Eisenstein saw D.W. Griffith's innovations as adaptations of literary conventions—Griffith's close-up was just a closer traditional theatrical viewpoint (Kaufman & Goldstein, 1976). Eisenstein and other Soviet filmmakers convinced of film's unique powers hoped to do more than adapt conventions from other media.

Eisenstein's ideas were influenced greatly by the environment of the Bolshevik Revolution, which brought new political and scientific approaches to the construction of art and aesthetics. Identifying with the artistic avant-garde, Eisenstein first gravitated to Meyerhold and constructivism. Meyerhold's aesthetics combined military drill and algebra, whereby the human body was seen almost as a robot. Taylorism, the study of a worker's physical movement and efficiency, also influenced Eisenstein and led him to a belief in the primacy of physiological gesture over psychological emotion. Consequently, Eisenstein saw himself as someone who constructs and assembles film like an engineer.

Freud and Pavlov further shaped Eisenstein's thinking, using Pavlov's notion of shock and stimulus in his approach to editing. A simple physiological approach to editing, this style excited emotions in the viewer, strengthening political consciousness or shaking ideological misconceptions. Furthermore, Wollen (1970) claims that Eisenstein, under the influence of Marxism, based his theory of editing on a somewhat vague notion of dialectics. Eisenstein's dialectical shot pattern of thesis-antithesis-synthesis can be seen as a question-answer structure, two shots in sequence call forth the question of what is the connection between two ideas represented by images (Katz, 1991). Thus, Eisenstein saw montage as collision and conflict. Under the influence of Kabuki theater later in his life, Eisenstein began to see montage as an activity of mental fusion or synthesis rather than conflict.

One of the major ambitions of Soviet filmmakers during the 1920s was to bring together politics with art and form with content. Sergei Eisenstein's theoretical writings on film have their genesis in the revolutionary environment of his time, combining a popularized version of Marxist dialectics with an impassioned vision of the new narrative art form's potential. Eisenstein was continually trying to apply a Hegelian/Marxist structure to his film problems. Consequently, a contradiction continually exists between the Marxist intellectual demands of his theoretical structure and the lyrical aesthetic interest in his film's physical world. Along with Griffith and the traditional theoreticians, Eisenstein views conflict as the basis for dramatic action: "Conflict as the fundamental principle for the existence of every art work and every art form" (Eisenstein, 1969, p.46). But Eisenstein takes the reasoning one step further in the Hegelian/Marxist manner by pointing to internal conflict between man's potential self and the political reality as the basis of existence: "It is art's task to make manifest the contradictions of 'Being'" (Eisenstein, 1969, p.46).

In a famous essay on Griffith and Dickens, Eisenstein applies a dialectical structure to an analytical comparison of Griffith's American cinema with the Soviet cinema. He claims that while Griffith makes important developments in parallel montage, he never brings the two editing lines together in a dialectical unity.

> ...the montage concept of Griffith, as a primarily parallel montage, appears to be a copy of his dualistic picture of the world, running in two parallel lines of poor and rich towards some hypothetical "reconciliation" where...the parallel lines would cross, that is, in that infinity, just as inaccessible as that "reconciliation" (Eisenstein, 1969, p.235).

For Eisenstein, editing is the collision of shots. A film doesn't develop along parallel storylines; rather it proceeds through conflict to create a higher unity. "For us the microcosm of montage had to be understood as a unity, which in the inner stress of contradictions is halved, in order to be re-assembled in a new unity" (Eisenstein, 1969, p. 235-6.)

Politically, the Marxist interpretation of reality is one of a capitalist world in continual stress from internal contradiction. While American film uses parallel montage to keep the lines of conflict separate, Eisenstein attempts to throw them together to create a new unity. Dialectical conflict is the primary principle in Eisenstein's theories of montage. The categorization of metric, rhythmic, tonal and overtonal montage are mere methods until positioned in conflict with one another in the film construct.

Eisenstein breaks the interaction of the conflicting elements into intervals that build dramatic tension through rhythm. Rhythm is created not by the building of shots, as in Pudovkin, but rather through the collision of independent shots. The close-up functions for Eisenstein as a montage cell that gives particular meaning to the larger thematic unity. The juxtaposition of these separate close-ups must finally reveal the ideological conception of the film as a whole, the overtonal montage. The philosophical background for this film theory comes in general from Hegel; the close-up, or montage cell, corresponds to the historical particulars, and the overtonal montage, to the general, objective progression of history. The manner in which Eisenstein views the relation between the montage cell and the overtonal whole roughly equates to Hegel's conception of the interactions of particular and universal truth: "For Truth is the Unity of the universal and subjective Will" (Hegel, 1956, p.39).

Finally, Eisenstein points to "intellectual montage" as the goal to be worked towards. Continuing his dialectical structural analysis of film, Eisenstein arrives at the Hegelian conception of "spirit." Through the constant collision of dialectical forces, society progresses to the spirit, or the intellectual cinema.

> The intellectual cinema will be that which resolves the conflict-juxtaposition of the physiological and intellectual overtones. Building a completely new form of cinematography the realization of revolution in the general history of culture: building a synthesis of science, art, and class militancy (Eisenstein, 1969, p. 83).

Another analogy Eisenstein makes between montage structures in a film and symphonic music composition comprises a perfect example of applying a Hegelian/Marxist system to film theory. All the historical particulars, or montage cells, progress towards a dialectical unity in the film whole. Each particular plays its part in the overall development of the narrative; each has responsibility for its individual role and its general movement. "Each montage piece has a dual purpose to continue the movement of its individual line and to build on the total line" (Eisenstein, 1969, p.76).

While melding revolutionary content into a revolutionary form is the background for Eisenstein's use of dialectics in his theoretical writings, Eisenstein at times acknowledges the need to conform style to content outside the dialectical concern. In rhythmic montage, abstract schemes of a sequence's length, for example, must depend upon the subject in the frame. "Here, in determining the lengths of the pieces, the content within the frame is a factor possessing equal rights to consideration. Abstract determination of the piece-length gives way to a flexible relationship of the actual lengths" (Eisenstein, 1969, p.73-74).

It is important to note that out of the five montage types that Eisenstein lists—metric, rhythmic, tonal, overtonal, and intellectual—only the last refers to a non-physical effect; all the other montage categories are specifically directed towards physical expression. The famous Odessa steps sequence's power resides in Eisenstein's artful combining of shots of physical motion into montage pieces of various rhythmic beats. The rigid mechanical movement of the soldier's marching feet is contrasted with the flowing, anarchic roll of the baby carriage. The shortening of the length of each montage piece as the sequence progresses, builds the tension

Figure 30: Eisenstein's Montage Theory

Rhythm	Collision of single shots
Order	Changes meaning
Re-order	Dramatic effect
Gesture	Physical representation
Dialectical	Question-answer
Collision	Conflict
Potential self	Political self
Parallel montage	Conflict separate
Close-up	Particular
Montage	General
Overtonal montage	Universal truth
Intellectual montage	Synthesis/spirit

and further develops the contrast between the soldiers marching and the carriage rolling. Clearly, the power of this sequence is in the physical dynamics of the rapid editing and the movement within the frame. Although this sequence's action might be termed "dialectical," it is not of an intellectual sort as in intellectual montages: "...overtones of an intellectual sort: i.e., conflict-juxtaposition of accompanying intellectual effects" (Eisenstein, 1969, p. 83).

In order to illustrate Eisenstein's approach to editing, I want to discuss a few examples from his films. In terms of authorship, the battle scenes in *Alexander Nevsky* are the most recognizably Eisensteinian. The story of a battle between the invading Germans and the defending Russians led by Nevsky would appear to be a good occasion for Eisenstein to apply a rational dialectical analysis; however, the real interest of the sequence lies elsewhere. The battle on the ice is held together by a construction of lines of fighting soldiers, in costumes of contrasting shades of white, gray, and black, advancing and retreating in geometrical patterns. As opposed to the beginning of the film, with the camera framed horizontally on pastoral cliffs and shorelines, the battle scene movements often occur in diagonal positions in the frame, thereby instilling a visually dynamic element in the action. The visual emphasis of the scene is on the physical conflict, not on any rational dialectics.

A large part of Eisenstein's *Strike* is a series of rapidly paced montage units of people swarming through the city streets. As in *Alexander Nevsky*, conflict in *Strike* is represented in a physical manner, not intellectually. Although the subject matter is class struggle, the film is not presented as a rational analysis of conflict. Eisenstein's ability to build tension and keep the audience's interest throughout the film essentially consists of shots of masses of people moving through the streets illustrates his dependence on the pure aesthetics of the montage construction. There is simply no rational way that Eisenstein could keep the viewer's attention for such a long amount of time with so little happening without making the physical movement itself an object of expression. Here we see where film aestheticism triumphs over film philosophy or political stance. In addition, Eisenstein inter-cuts shots of workers running through the frame with a still shot of the factory whistle blowing. On a literal level, the whistle announces the beginning of the strike, but in terms of the sequence rhythmic structure, the still shot acts as a physical jolt to the viewer. This sequence represents the way Eisenstein's editing often works in his films: there is a rational thematic concern and a cinematic meaning system operating simultaneously. The film creates its effect on the audience through its physical expressiveness.

Visual metaphor is another technique Eisenstein employed. In *October*, he compares Kerensky to a peacock as a straight psychological analogy to the historical figure's narcissism. Later, Eisenstein criticized such "literary" metaphors as outside the sequence's visual dynamics. Speaking of metaphor, Eisenstein says: "Such a means may decay pathologically if the essential viewpoint-emotional dynamization of the subject is lost. As soon as the filmmaker loses sight of this essence the means ossifies into lifeless literary symbolism" (Eisenstein, 1969, p.58).

Eisenstein points here to his reliance upon physical correlatives in creating successful visual metaphors: visual metaphors work through physical likeness or through movements of the objects in the two separate frames. The traditional view of Eisenstein's use of metaphor as being a kind of psychological "this is this" reflects a misunderstanding of his narrative expression's purely physical nature.

In *Strike*, Eisenstein creates a cross montage between the killing of workers and the butchering of a bull. He emphasizes the physical correlation of the two events by cutting between a close-up on the bull's struggling legs and fleeing workers running diagonally down a hill. Eisenstein comments on this scene himself: "As a matter of fact, homogeneity of gesture plays an important part in this case in achieving the effect—both the movement of the dynamic gesture within the frame, and the static gesture dividing the frame graphically" (Eisenstein, 1969, pp. 57-8).

While this sequence has a rational connotation, it is primarily effective on a visual/physical level. In *Strike*, the great number of visual metaphors creates a non-linear narrative that proceeds more associatively than rationally. There are a number of dissolves between close-ups of various animals and characters in the film—comparing, for example, the spy with a fox. The large wheel gears of the factory is another recurring metaphor: three workers are shot through a spinning gear; a worker's meeting is set at a yard with an enormous stack of old gears; and a medium-shot of police spraying a man with water is inter-cut with a shot of the gears spinning at the factory. All the metaphors rely on the compared objects' physical expressiveness in order to work. More than being intellectual correlatives, the visual metaphors are meant to make physical comparisons.

Ultimately, Eisenstein was unable to make his art conform to his political/philosophical thinking. His beginnings with Meyerhold and biomechanics taught him to look towards the body as a means of expression—that through a series of body assaults, the viewer's mental state could be changed—and eventually this leads Eisenstein into an aesthetic fascination with the physical world that is antithetical to the dialectical system. Eisenstein's desire to film Joyce's *Ulysses* is symptomatic of his preoccupation with the subjective, particular, and physical aspects of life.

In the Hegelian scheme, Eisenstein wrote about film in relation to the general objective conception, but often made films with an interest in particular and subjective truth. In a teaching session, Eisenstein quotes Hegel's aesthetics:

> The only great method—is to have no method. Thus, in your own creative work, you must start every time from the story, the art form, the political idea and world outlook. If you want to have a method, then clearly understand the task and seek accurately and concretely for its solution (Nizhny, 1962, p. 110).

Here, Eisenstein clearly expresses his primary interest in film production rather than the method of approach. Eisenstein is known for his editing techniques, but the graphic composition within each montage cell is also of great interest to him. A formal composing of volume, light, and spatial relations runs throughout his films. In both the montage sequence and the individual cell, the volume, weight, and movement of the objects in the frame are more the proper subjects of the film than the rational structure.

What can we learn from Eisenstein that has practical application to designing computer educational environments? Eisenstein's theoretical writing and his editing techniques raise many important issues for education because he focused on the transmission of ideology. He asked how meaning is constructed through a sequence of images—a key question for educators. His approach using dialectical conflict is one that can be useful in educational environments particularly as it connects to question-answer formats. By juxtaposing concepts, ideas and images in various orders, a high-level understanding might be achieved. Finally, Eisenstein's employment of physical metaphors is often powerful in his films and may be used in educational environments for effect.

NARRATIVE VERSUS NON-NARRATIVE

If learning is increased through the use of narrative, how do you address the question of illusionistic narratives that lead to passive and uncritical learning? According to Erickson, Rossi and Rossi (1976), learning can involve a trance state. Narrative films consciously work to create this kind of trance state to involve viewers and retain their interest. This is a primary advantage of narrative used in educational settings: it captures the learner's attention and may even speak directly to the unconscious.

However, is this desirable? In the literature on cinema, narrative film is considered illusionistic—not something that brings forth critical thinking. In the previous chapter we saw that documentary film, Brecht, and the Surrealist Movement address this question through multiple techniques, many of which might be applied to narrative in educational environments. The strategies of these approaches are explicitly non-narrative or anti-narrative in a Brechtian sense (we saw in the previous chapter that Bertolt Brecht argued for interrupting dramatic performance in order to force the audience to think critically and be aware of the artificiality of the performance). Narrative devices in documentary film such as voice-over, editing approaches, and characterization have long been a subject of debate because they might be abused to promote political points of view.

In fiction film, the Italian Neo-Realists and the French New Wave movements explicitly attempted, in various degrees, to subvert traditional Hollywood narrative film patterns in order to allow (or force) the audience to think more critically. This subversion was accomplished by such neo-realists filmmakers as Rossellini, DeSica, and Antonioni with the use of wide angles, deep focus, and long takes. The New Wave directors, including Godard and Truffaut, used a more Brechtian technique of self-reflexive story content, jump cuts, and excessive camera movements to make the audience aware of storytelling's artifice. The history of film theory and practice in fact shows a clear division and conflict between illusionistic Hollywood-style narratives and non-narrative or non-illusionistic film.

Janet Murray (1997) points out that computer fictions tend to test the border of real and unreal. This question of how the viewer regards the representation's realness in a computer environment is of primary importance when constructing software. In film criticism, the issue of realness is defined in discussions of narrative as involving illusionistic narratives and passive viewers. For some filmmakers, the desire to avoid this illusionistic tendency has led them to promote the creation of non-narrative films. Two theorists' philosophies reflect different aspects of this position. A proponent of non-narrative structure, Peter Gidal (1989) intends through his structural/materialist theoretical position to distance the viewer from the work by making him or her continually aware of the cinematic apparatus. Through repetition, minimization of image content, and paralleling viewing time with produc-

Figure 32: Narrative Versus Non-Narrative

Narrative	Non-narrative
Content	Form
Passive viewer	Active viewer
Uncritical	Critical
Illusion	Visible process
Voice-over	Live sound
Characters	Real people
Narrow camera angle	Wide angle
Shallow focus	Deep focus
Short segments	Long takes
Congruent editing	Jump cuts
Motivated camera movement	Excessive camera movement
No reference to space out of frame	Testing borders
New shots	Repetition
Full image content	Minimal image content
Identification with characters	Outside character positioning
Dominant film codes	Subversion of codes

tion time, Gidal wants to shift attention to the film's formal construction. On the other side, Noel Burch (1981) believes form should express content and that the advancement of film techniques will expand the medium's content. Both theorists take formalist positions, but Gidal excludes everything but form as a film subject, while Burch simply wants to expand the film's narrative structure through various techniques that he sees as being repressed by the dominant codes of traditional film. Gidal and Burch's contrasting theories relate directly to a discussion of educational software because the assertion of realness is particularly important in educational products. As we saw earlier, while narrative has distinct advantages for educational purposes, it also brings dangers.

Peter Gidal defines structural/materialism as film that tries to demystify the viewing process by continually destroying the illusion of narrativity. The major problem in his theory is making the production process visible to the viewer: "...that of making visible the procedure, presenting such as opposed to using it" (Gidal, 1989, p.189). For Gidal, awareness of the production process when viewing film is essential to anti-illusionism. A linear narrative line doesn't structure the film; rather, it is the mind of the viewer attempting to organize and analyze the particular images; the narrativization agent is intended to shift from the interior of the film work to the mind of the viewer. The structuring aspects and the attempt to decipher the structure and anticipate/recorrect it, to clarify and analyze the production process of the specific image at any moment, are the root concern of structural/materialist. Applied to educational environments, this would mean a constant attention to how content is delivered.

A structural/materialist work does not present the viewer with a narrative whole; it evokes in the viewer's mind a need to interject meaning into the film. Where traditional narrative film lays out the story information in a clear line and thus creates a passive viewer, the structural/materialist film presents relativistic moments that the viewer must confront and actively participant in. Minimization of image content within the frame is one device Gidal suggests to shift viewer interest from film's content to the form of film. Content within the frame serves simply as a subject to be manipulated by the film structure. In addition, for Gidal, in contrast to the short takes of traditional narrative film, the length of the take in production time is closely related to the final screen time. In trying to break the dominant codes of narrative that suppress the "real" time of production by restricting the screen time, the structural/materialist film tries to bring the production time and the viewing time into a closer relation. Michael Snow's famous non-narrative film *Wavelength* is an example of Gidal's conception of bringing production and screen time closer together. Repetition is another film device that makes use of time to bring out the structure of the film work. Used over a substantial length of film time, repetition shifts the viewer's attention from the dominant narrative codes to the film construct. An example of this is *Serene Velocity* by Gehr, where the rapid zooming back and forth in a hallway repeated over a length of time eventually breaks down the viewer's referential connotations of "hallway" and shifts the interest to the moving vertical lines and the changing perspectives.

Like Gidal, Noel Burch (1981) approaches film through analyzing its formal structure. Both theorists believe that "bad film" is made through the simple, unthinking application of the dominant codes of narrative to specific works. For Gidal, it is necessary to do away with narrative completely to make authentic films that rely solely on film form. Burch promotes usurping dominant codes to create an "open" system of narration that can freely expand formal devices in conjunction with the evolving film content, and thereby develop an authentic cinema. Burch advocates an in- and out-of-frame dialectical concept as an innovative manipulation of space. He cites Renoir's *Nana* for its progressive use of characters' entrances and exits in and out of the frame, as a technique for making use of the space behind the camera. Characters looking off screen, framing characters so that portions of their body are out-of-frame, and using empty frames all draw attention to the off-screen space and thus expand the potential area of viewer interest beyond traditional cinema's 180 degrees.

For Burch, a shot's temporal duration is conditioned by its readableness, its "legibility." His notion of a progressive use of shot duration is to alternate the degree of difficulty in understanding an image with the length of the shot. Burch wants to develop tension through combining shots that are too short to read with shots that are too long and boring. By bringing together the shots of various perceptible lengths in an artistic manner, visual rhythm can become more complex. To Gidal, Burch's use of time in film simply constitutes an expansion of the film illusion. For Gidal, film time must be used through long-takes and repetition to break down interest in the film narrative, not to expand it.

Burch claims that accepted masterpieces of cinema are films in which the dominant codes are subverted, in which structure questions itself. Burch rewrites film history and its achievements in relation to the degree of subversion by the film of the dominant codes of representation and narration. In this scheme, *The Cabinet of Doctor Caligari* is the first great film because it is the first self-reflexive film work: that is, it avoids the ideological stance through its deconstruction of traditional cinematic codes. Burch points to various cinematic techniques used in *The Cabinet of Doctor Caligari* that work against or "deconstruct" the dominant narrative codes. For example, the use of theatrical frontal shots and dissolves are applied to a series of shots that appear to be narratively continuous, giving each shot autonomy within the narrative construct; in this way, each shot works against the invisible flow of the narrative. Moreover, the illusion of spatial depth is opposed by the physical flatness of the screen. The expressionistic painted sets exaggerate perspective on an obviously two-dimensional surface, and the actors walk through the frame in a way that emphasizes the depth of the area. In *The Cabinet of Doctor Caligari*, the set dominates the narration and thus opposes the traditional narrative line, which relies on characters for plot motivation. Finally, Burch cites the narrative's ambiguity, directed by a madman, as a subversion of the storyline's reality.

Gidal attacks the practice of "reading into" the film work as he claims Burch does in analyzing *The Cabinet of Doctor Caligari*. In contrast to Burch's ideas on Dreyer, Gidal downplays the filmmaker's innovation in narrative as a simple transference of viewer identification from the psychological/emotional to the psychological/rational. For Gidal, Dreyer keeps the same identification process of traditional cinema but changes the character's motivation and story interpretation into a rational scheme instead of an emotional one. The problem with Burch's deconstruction of film classics, as Gidal points out, is that he projects his own theoretical bias onto the films to such an extreme degree that it becomes forced. Rather conservative, mainstream films become vehicles for the radical undermining of dominant cinematic codes. Burch goes as far as claiming that it is precisely the subversion of these codes that makes such films great. Although these films do make use of innovative devices, their major narrative lines are generally developed using the dominant film codes of their respective periods.

What are the implications of Gidal and Burch's theoretical work for education? Burch sees the development of formal aspects as finally working with the narrative content in a reciprocal kind of determination. For both Burch and Gidal, form is content. One would imagine that both Burch and Gidal would insist on educational media focusing especially hard on form. But for Burch, form should not devalue content as in Gidal, rather it should more completely express content: "That form is content, and that content can create form" (Burch, 1973, p.88). Burch takes a more traditional approach to form, one that would allow educational software designers to focus on form while still providing content. For Gidal, narrative is essentially illusionist because it denies viewer consciousness of the film process. Narrative film doesn't allow for the proper distancing of the viewer and thereby forces him into a passive role. The important point for educational software designers here is that there is a tendency in film for viewers to position themselves in a passive role somewhat automatically. For Gidal, an actualized film would have to be non-narrative, not simply a film with free or open codes. Consequently, Gidal would most likely broadly criticize attempts to put educational materials into the form of narrative, as he sees narrative as working in media to make a viewer passive and simply accept in an unthinking manner the dominate codes that are presented.

In terms of creating educational environments using media, both critics emphasize the importance of viewer knowledge of structure and making viewers aware of form, in order to create an intelligent viewer and user of media.

Laura Mulvey (1996), another influential Marxist theorist who attacks narrative film, argues that Hollywood uses a Detroit-like system of production with an elaborate style characterized by the erasure of its own mechanics. These films consciously attempt to hide the means of production in the film product. Consequently, Hollywood cinema is a symptom of cultural, economic, and technical fetishism, according to Mulvey. Fetishism attaches itself to products of labor when they are produced as commodities. Feminist film theorists such as Mulvey argue that cinema finds its most perfect fetishistic object in the image of the woman. Consequently, during the 1970s feminist film critics pushed towards the defetishisation of the film medium. The contribution of feminists film theory to this discussion about the use of images in educational environments is important because it emphasizes the danger of illusionistic approaches that can easily lead to turning the educational program into a commodity.

Clearly the debate about narrative and non-narrative approaches is important to designers of educational software because it reveals how specific media techniques lead to significant positioning of the learner. We can see from this review of the debate on narrative versus non-narrative film that there are legitimate concerns about the use of the film for educational purposes. Another aspect of narrative is connected to how sound is employed.

SOUND

One large difference between feature films and computer software is the use of sound. While in the last 20 years in American cinema directors such as Speilberg and Lucas have focused on the narrative uses of sound, most computer applications vastly underuse it. Computers thus far heavily emphasize the visual message over the auditory. When educational software does employ sound, it often involves a voice-over narrator and adopts a traditional documentary film approach.

The research literature shows that sound in film has received far less attention than visual aspects of the medium. Altman (1992) claims that because photography used a familiar approach, the public readily accepted it. Likewise, sound in movies first conformed to established uses in records, radio, theater, and public address and operated at natural levels. Soon, however, sound evolved to emphasize the narrative and dialogue. Originally adapted from radio programs, early sound films tried to represent space through sound. After the introduction of sound, a period of conflict over approaches to use began with attention devoted either to space or speech. Mitry (1997) argues that because film can record human voices it was naturally pushed more and more towards realism. However, according to Altman, sound is better understood as representation than reproduction.

Altman (1992) argues that film should be viewed as an event instead of text in order to understand that the use of sound involves three-dimensionality, materiality, instability, diffusion, and interchange. Documentary sound is quite different from fiction film sound, in that speech may be inaudible, sound spaces differ radically between scenes, and there may be a general lack of clarity. Documentaries have paid special attention to the quality and appeal of their musical scores (Karlin, 1994), many of which have been composed by concert hall masters such as Morton Gould, Robert Russell Bennett, Oscar Levant, Paul Creston, Gail Kubik, Mark Blitzstein, and others. At the other documentary extreme, newsreels used canned music recorded as library music then used as needed.

The 2001 survey found that respondents saw the importance of narration over realistic sound levels (Figure 33). In response to the statement, "Clarity of voice-over narration is more important than realistic sound levels," students agreed or strongly agreed 69.7% of the time.

One might conclude from these data that the students again prefer passive learning experiences over the presentation of source material in the form of live sound that needs to be interpreted in some manner. On the other hand, most of the students indicated that background music was not important (Figure 34). In response to the statement, "I would prefer some background music for distance learning programs," 79.4% indicated that they either disagree or strongly disagree.

As noted above, voice-over narration in educational software often dominates sound, because narration typically organizes the fractured visual content. Thus, in

Figure 33: Voice-Over Narration Importance (Questions 1 & 13)

Delivery format * Clarity of voice-over narration is more important than realistic sound levels. Crosstabulation

			Clarity of voice-over narration is more important than realistic sound levels.				Total
			strongly agree	agree	disagree	strongly disagree	
Delivery format	computer-based	Count	2	20	12	1	35
		% within Delivery format	5.7%	57.1%	34.3%	2.9%	100.0%
	videotape	Count	5	31	12		48
		% within Delivery format	10.4%	64.6%	25.0%		100.0%
	correspondence	Count		1	1		2
		% within Delivery format		50.0%	50.0%		100.0%
	other	Count		1			1
		% within Delivery format		100.0%			100.0%
Total		Count	7	53	25	1	86
		% within Delivery format	8.1%	61.6%	29.1%	1.2%	100.0%

some ways, the scripted narration, not the images, are often the strongest carrier of the lesson content and therefore the strongest aspect of the teaching film's style. Nevertheless, as noted in Chapter Eight, documentary film theory warns that voice-over narration works against critical thinking and is therefore not a good learning approach.

Figure 34: Background Music Preference (Questions 1 & 14)

Delivery format * I would prefer some background music for distance learning programs. Crosstabulation

			\multicolumn{4}{c	}{I would prefer some background music for distance learning programs.}	
			strongly agree	agree	disagree	strongly disagree	Total
Delivery format	computer-based	Count	1	8	26	11	46
		% within Delivery format	2.2%	17.4%	56.5%	23.9%	100.0%
	videotape	Count		10	34	3	47
		% within Delivery format		21.3%	72.3%	6.4%	100.0%
	correspondence	Count		1	1	1	3
		% within Delivery format		33.3%	33.3%	33.3%	100.0%
	other	Count			1		1
		% within Delivery format			100.0%		100.0%
Total		Count	1	19	62	15	97
		% within Delivery format	1.0%	19.6%	63.9%	15.5%	100.0%

CONCLUSION

In the elaborate and often jargonistic film theory and criticism literature reviewed in this chapter, we found a concentration on editing principles and conventions. Henderson (1976) argues that there are two principle types of film theory that relate to editing strategies: part/whole and theories of relation to the real. Eisenstein and Pudovkin exemplify the part/whole category, while Bazin and Kracauer are in the real camp. According to Henderson, Eisenstein is weak on formal wholes because of his commitment to the part-complex (the sequence) as the aesthetic center and his concern for control; Bazin's weakness is that he does not go beyond the real. In terms of educational applications of these theories, Eisenstein's approach is useful in emphasizing rhetorical approaches, while Bazin's is more likely to create active learners. According to Mitry (1997), cinema is the art of association and suggestion—essentially editing functions. Likewise, making connections or associations among concepts and ideas is an important function of education. While suggestion is primarily an artistic activity, it might be fruitfully engaged in more sophisticated educational approaches. Clearly, those developing strategies for the use of images in computer-based educational software need to direct attention to editing. In fact, it may be the most important issue in constructing computer-based educational environments.

In the next chapter, I look more deeply within the mind of the computer user to explore subjective psychological states and reactions to media through point of view. I also delve into how viewing media is like dreaming, and how wishes function in media.

Chapter XI

Subjectivity, Point of View, and Dreaming

In this chapter I continue to concentrate on media and the mind of the computer user and now turn to a psychoanalytic analysis of the viewing experience as it may apply to learners. Extending the focus of the second part of this book, I attempt here to understand psychological states and reactions to media through point of view, how viewing media is like dreaming, and the function of identification, as well as perceive overall how personal fantasies function in media.

POINT OF VIEW AND SUBJECTIVITY

Implied point of view is a main structuring device used in film editing, and also works to position a learner when using computer media. A subjective positioning of the viewer through editing has a long and complex history in film theory and criticism. Subjectivity is another way to talk about point of view. Hedges (1991) points out that the rise of the use of the camera occurred at the same time that psychoanalysis came to prominence. The end of the 19th century marked a turn towards subjectivity with the concentration on both trying to understand the individual's mind and in using a new photographic device that captured specific points of view. Camera shots are always taken from both a real and implied point of view. The photographer takes the pictures by looking through a viewfinder, and depending on the camera, the resulting photograph is a record of that point of view. Additionally, by using various techniques, the photograph can imply points of view other than that of the photographer. Thus the point of view taken through film, video, still photography, text, and software is of central importance in the design of educational environments. We will see in this chapter that the research literature on point of view in film is particularly rich and has great application to educational software.

One of the primarily responsibilities of a film director is to control point of view by manipulating narrative logic, eye contact, and shot size. During the primitive years of film, the emphasis was on movement for movement's sake without a lot of consideration about narrative point of view (Brownlow, 1968). Silent film based itself on theatre for dramatic structure. Talkies used theatrical style dialogue and stories and as a result often positioned the viewers in the place of a stage audience (Mitry, 1997). However, it was quickly discovered that in editing, the most powerful cuing device is the sightline of an actor in close up (Katz, 1991). This sightline implies point of view and directs the viewer to understand from whose position subsequent shots should be understood. So even in early silent film, it was common to show a picture of an actor looking off-camera followed by a shot of what the actor was looking at. This example probably indicates the simplest form of the use of point of view positioning, but there are many more much complicated and subtle forms. Indeed, Katz (1991) points out that there are degrees of subjectivity in identification for any shot in a film. In the same way, educational computer environments often present various degrees of subjective experiences for the learner. Two examples of this show a lesser and stronger degree of subjectivity. In the *Madeline 2nd Grade* program, the user watches over the little girl's shoulder as she solves math problems with chalk on a sidewalk. A stronger form of identification is found in *Mavis Beacon*, an application that teaches typing skills by actually showing a real-time animation of the user's fingers as they move on the keyboard.

In the film criticism literature, authors often talk about "open" and "closed" compositions. Open means a documentary style with less and more transparent points of view, while closed means that the shot appears to be designed and arranged in a conscious effort to express the author's point of view. In this way, simple decisions such as the closeness of a shot to the subject imply degree of open or closeness. The tighter the shot, the more the intimacy with a subject is implied, making the audience identify more with the subject or with his or her specific point of view.

Christian Metz (1982) asserts that cinema lags behind verbal language in ability to portray the subjective viewpoint because of its phenomenological relationship to the world in physically making an impression of reality. There is a problem in film with finding physical correspondences for mental states. On a political level, Hedges (1991) points out that subjective film points of view tend to reinforce dominant

Figure 35: Point of View

Open	Closed
Documentary	Fiction
Clear	Hidden
Random	Designed
Multi-gender	Male
Multiple cultural positions	Dominant cultural position
Real locations	Expressive sets
Real or minimal lighting	Expressive lighting
Multiple or non-specific	Author-controlled

cultural positions. For instance, most films represent a man's point of view and rarely identify the point of view of a woman character even in contemporary filmmaking.

The most famous and earliest use of subjectivity in film was in the German Expressionist films. Connected to the Expressionist Movement in art and theater, the film movement focused on the expression of internal psychological states of

Figure 36: German Expressionism—The Cabinet of Dr. Caligari (1919)

characters. As discussed earlier in connection with Bertol Brecht, these powerful films used techniques such as extreme and exaggerated set design and camera angles.

Additionally, chiaroscuro lighting with strong shadows and the use of close-ups were commonly used in these films. These films and the filmmakers who emigrated to America had a very strong influence on Hollywood, and the techniques expressing subjective states are still used to good effect in contemporary film. Later, in the early 1960s, filmmakers began to assert themselves as cinematic authors, attempting to make more personal films that expressed their own point of view. In particular, New Wave filmmakers such as Jean-Luc Godard, Francois Truffaut, and Eric Rohmer, many starting as critics for *Cahiers du Cinema* with general notion of directors as film authors, began to make extremely personal and subjective films.

A central question that needs to be asked about educational media is how point of view or subjectivity is represented. What position for the audience is either explicitly or implicitly represented? For educational applications, how can the techniques of film to represent point of view and subjective stages be used to increase learning? It seems to me that such debate has great relevance to the design of computer interfaces for educational purposes. Might not it be important to incorporate cinematic techniques into interface designs that make the learner aware of narrative point of view and the artifice of meaning construction? How is this done in a way that is most productive for the individual learner? After all, motion pictures are made for general audiences, whereas computers are personal. This represents both a challenge and a great advantage. One way to look at point of view in computer environments is to think of the notion of "voices." Brooks (1999) points out that although the computer can accommodate many voices through its search and storage capabilities, it does not automatically make sense of the voices. He proposes a meta-linear narrative strategy for fictional works, a collection of small related story pieces designed to be arranged in many ways, to tell many linear stories from different points of view, with the aid of a computer story engine. For instance, this approach for a history application about the civil war might allow a user to view the content from various northern, southern, or slave points of view. Meta-linear narrative addresses the problems of authoring and presenting multiple points of view.

Abrahamson (1998) argues that rather than presenting course content in an objective fashion, storytelling allows the teacher to insert his or her own point of view in a conscious manner. He claims that this technique makes the teacher's viewpoint obvious. As leading film critics argue, films come to the viewer as either impressions of reality or as dreams and are more convincing that other symbolic systems. Film has been extensively used in propaganda for this very reason ever since the use of film in Nazi Germany as epitomized by Leni Riefenstahl's *Truimph of the Will* (1935). What are the educational and political ramifications when using powerful film techniques in computer educational environments? In particular, how does the likeness of watching moving images to dreaming affect computer-based students?

MEDIA VIEWING AND DREAMING

> ...the true enjoyment of literature proceeds from the release of tension in our minds. Perhaps much that brings about this result consists in the writer's putting us into a position in which we can enjoy our own daydreams without reproach or shame (Freud, 1963, p.43).

One way to think about media point of view is to concentrate on how the viewer/learner perceives his/her experience. In recent years, understanding film viewing as dreaming has become a common approach in film criticism. The shift that occurred was to think about what happens inside the viewer psychologically. What happens to viewers when watching a film? Consider the following: imagine yourself in a warm, dark room, cavernous, yet strangely secure, pushing back into a soft chair. You are surrounded by an ocean of faces bright with reflected light, yet you feel alone, unwatched. In the darkness you doze off into a shared dreamscape populated with detectives, cowboys, bimbos, monsters, along with life-like mothers, fathers, and children. Images the size of a building flicker in front of you; they come in a speckled stream of light from a box behind. It's a convincing shadowplay, one with substance. Everything on the screen is in the eternal present tense, always occurring now. Although it comes from the box behind for all to see, the enormous shadow is your first-hand experience, your wish, your dream. Watching a film is like dreaming.

We've all had the experience of an informal discussion with a group of friends about the latest films and observed how even the most "intelligent" opinions are incredibly subjective. People with very similar backgrounds—who generally agree about politics, fashion, music, and books—often disagree when it comes to their taste in films. The reason for this is that films, especially Hollywood-style films, appeal to us on a very subjective, uncritical level—we experience them as dreams.

Ever imagine what it would be like to be a man accepted inside the world of women (*Some Like it Hot*)? Or as a child, left on your own, to outsmart two adults (*Home Alone*)? Or for women, image that you turn the tables on the dominant male world (*Thelma and Louise*)? A film is a personal dream experience for each viewer, and individuals judge the experience based on the degree to which the film conforms to their own private fantasy structures. According to Freud, all dreams are wishes of one kind or another. The primary conscious or unconscious question for any filmgoer when deciding on which movie to see on a Saturday night is: "Is this a dream I want to have?"

Film criticism in the 1980s turned to the use of psychoanalysis in order to help understand the audience, and how an audience perceives the sounds and images it encounters in a movie theater. In this analysis, a film is like a dream, with each scene built upon public and private wishes. While scholars often don't discuss the audience, it is crucial that we do so, particularly when looking at educational applications of media. Film and television media have an incredibly strong influence on audiences, and their appeal is psychological in a very immediate way.

While it may seem odd to think about dreaming as a strategy for learning, there is a very deep literature connecting film viewing to dreaming. Additionally, there is a growing body of research literature connecting dreaming to the experience of computer users. Sherry Turkle (1995) sees the computer as a new medium on which to project fantasies, and analyzes a trend towards people wanting to substitute computer images for representations of the real, seeking out the subjectivity that computer simulations offer. Others concerned with computer interface design and interested in the psychological processes of the computer user have looked at dreams as an organizing principle. Davenport, Agamanolis, Bradley and Sparacino (1997) write about creating a "Dream Machine" that uses the rich language of dreams for a networked audience. Durbridge and Stratfold (1996) created "Art Explorer," a multimedia system based on the analogy of dreams. It uses the learner's experiences to build a sequence of personalized "discrete happenings" using an episodic narrative structure.

American films are known for their simplicity, for the economic and direct way they tell a story. Why do we want to see such simple stories? Why is it that certain stories appeal to us and others don't? When we look at films as dreams we can begin to make sense of the American preoccupation with particular types of subjects. There are two fantasy structures that come into play for a film audience: an individual/personal fantasy structure, and a collective/social fantasy structure. Obviously, the collective/social fantasy structure is what many filmmakers try to appeal to because it means bigger receipts at the box office. For educational software designers, it can mean a more accessible approach to involve learners.

In *The Imaginary Signifier* (1982), Christian Metz expresses a psychoanalytic perspective on the interpretation of films as dreams with a focus on the film viewer. A leading film theorist, Metz argues that films come to viewers as dreams through the physical immobility of the spectator, the darkness of the theater, and the phenomenological nature of film that he describes as both absence and presence. The unique position of cinema lies in this dual character of its signifier as both perceptual wealth and unreality. Metz focuses on the psychology of the film viewer

Figure 37: Media Viewing As Dreaming

Learner	Wish or promise
Real	Dream
Viewing situation	Darkness/screen size/distractions
Voyeur	Perceptional wealth
Unattainable wish	Absence/presence
Desire	Lost imaginary state
Repetition/displacement	Plot
Neurotic	Repressed/non-repressed story elements
Lure	Disavowal of death

and sees film as a mirror showing a psychological lack that draws the viewer in. The fact that film presents an image of physical presence without real presence of objects creates desire in the viewer. For Metz, films are wishes, wishes that can never be fulfilled. Metz's viewpoint is not necessarily in conflict with the cognitive position in as much it fits within a broader notion of culturally derived meaning making. What he potentially adds to this larger discussion is that if viewers receive films as dreams, narratives might best be structured to take advantage of this predisposition. For Metz, there is a two-fold process involved in watching a film. First, the film presents itself as a particularly rich perceptual event that tends to deny its existence as fiction. Secondly, the conditions of the film-going experience are such that the viewer is particularly receptive to its spectacle.

Metz adopts Melanie Klein's notion of the splitting of objects of fantasy into "good" and "bad" objects: good objects conform to the individual's fantasy wish, and bad objects avoid or repress that wish. Film, in relation to fantasy structures, is both too certain and not certain enough. A film can never conform itself as exactly to the individual's wishes as a dream can; the dream is more exact and consistent in its dramatization of desire. The very particularness of the film medium that supplies an impression of reality can also be the cause of spectator rejection. The specifics of each individual spectator's fantasy structure often doesn't fit the fiction presented in a film; the film is then a bad object. This concept goes far in explaining why film viewers seem to be particularly irrational in their likes and dislikes. The notion of disappointed fantasy is noticeably evident in film adaptations of novels. The reader of the novel generally has already clothed the story in his or her own fantasy specifics, and therefore, often finds the director's conception inauthentic.

Metz connects cinema to fantasy structures by analyzing the desire to see, what he calls "the scopic drive," in the film viewer. He locates the scopic drive by applying Jacques Lacan's, the French philosopher, notion of the individual as founded on an essential misrecognition of the self in terms of an original unity with the mother. For Lacan, the "Mirror Stage" represents the primal scene of self-alienation; The Imaginary is this first misrecognition by the individual in understanding the self in terms of the unity with the Mother. This misrecognition of the self in terms of unity is broken by the intervention of the Father representing the Symbolic: language and the entire cultural meaning system. Forever after the entrance of the Father (the Symbolic), the individual will always yearn back to the original unity (the Imaginary). The scopic drive is an infinite pursuit of this always absent self.

For Freud, the scopophilic drive is always a desire to look at other objects as if they are part of one's own body. Film feeds this scopic drive by offering the illusion of an object's presence, yet always keeping it unattainable—every film is in this way a broken promise. It is precisely this absence/presence that fixes scopophilia to film; there can be no voyeurism connected with objects that are obtainable, for they immediately extinguish the desire. It is in the presentation of physical objects that fetishism intervenes. Desire is constituted on the lack of its object; thus film provides the perfect device for the pursuit of the unattainable.

The essential characteristic of the Hollywood film is that it is a specific type of wish fulfillment discourse that passes itself off as a story. Each film denies that it lacks anything, or that it meets any desire of the film viewer, except in terms of a generalized "entertainment." The film presents itself as a story that fulfills a wish that was never requested. For Metz, this is the moment at which the cinema reveals itself as a profoundly ideological medium. The film viewer, as a voyeur, is a vacant receiver—he accepts film images in an affective film as if they were his own creation.

In order to understand psychoanalytic interpretations of film narratives, it is useful to look at an example. Furthermore, I present here different philosophical viewpoints in connection with psychoanalytic perspectives to get a full sense of the issues involved. Rodowick (1980) traces the course of desire in the textual system of *The Pirate* (director: Minnelli, 1948). He sees the film being organized around the "structuring absence" of the romantic image of the pirate. Rodowick follows the circulation of this essential lack through the textual system in terms of repression, delay, and displacement. The economy of desire in *The Pirate* is preserved in a continual repetition of a fixing of desire to an object, and then a rejection.

Specifically, desire is restricted by the gaze of Manuela, the main character, who constantly misdirects it. In the look, desire finds form. It is in this way that the film shifts from the position of Manuela to the traditional male point of view and Judy Garland becomes the object of the gaze, instead of its source. For Rodowick, *The Pirate* is interesting because it exemplifies the manner whereby fiction film arranges women's desire so that it is expressed in her as a performer, but not for her as a viewer. The desire for *The Pirate* is the same lack that Metz proposes in his phenomenology of the cinema: desire is structured around the lost Imaginary state. The articulation of the lack in terms of *The Pirate* provides the logic of repression in the text. The circulation of desire in the text leads back to the viewer's subconscious, which is founded on an ego construction of loss. Accordingly, each film text is an articulation of desire in which repetition and displacement serve as temporary representatives of this loss.

For Rodowick, the artistic work serves as a kind of dramatization of desire. Jean-Francois Lyotard (1992), the French scholar, counters this analysis by arguing that drives are not so easily systemized. Using the metaphor of a volcano, he proposes that Freud's conception of drives was a block of often contradictory forces. His reading of Freud claims that drives express themselves in ways that break the rules of logic, and therefore cannot be systematized. Lyotard attacks the psychoanalytic analysis of artistic practice for its oversimplification of the process of conversion of drives into psychic representation. He argues that the artistic event cannot be understood as a readable discourse of the unconscious. Instead of analyzing art works in terms of would-be symptoms, Lyotard suggests interpreting them as presented perspectives of reality. He sees art as an opportunity for expanding the individual possibilities for experiencing reality.

On the other hand, Philippe Lacoue-Labarthe (1979) criticizes Lyotard for assuming that a reality exists outside of representation. The proliferation of represented perspectives that he notes as the primary interest of artistic works is based on a notion of the possibility of primary knowledge of the "Real." Lyotard is in some sense proposing the concept of an un-alienated subject. Lacoue-Labarthe applies Nietzsche to Freud in proposing that the enjoyment of art is masochistic. Whereas the early Greek drama deals solely with equally conscious oppositions, with the development of modern art forms, the conflict becomes centered on repressed/non-repressed inter-play. Thus for Freud, the viewer who enjoys modern forms of art must be a neurotic; a normal viewer would simply supress the repressed elements in the drama and thereby defeat the drama. This point is further bolted down by the need for identification within the work for the viewer/reader of the repressed elements in the text. In this way, the viewer must give equal value to the repressed elements of himself and thereby be neurotic. Finally, Lacoue-Labarthe takes the absence/lack of Metz and Rodowick one step further and identifies it as death. For him the split of representation is between desire and death; death submits to representation. In tragedy we play out our constant identification with death; it is that which destroys all "presence" in signification. The lure of art then, is in its disavowal of death.

The theoretical writings of Metz, Rodowick, Lyotard, and Labarthe all compliment each other in further examining how viewers/learners respond to media in computer learning environments, a central concern of this book. We have seen how in film criticism one of the important shifts that occurred at the end of the 20th century was a concentration on the cognitive processes of the viewer watching films. This shift also needs to occur in considering educational software. When one looks at how the learner perceives educational images, psychoanalytic functions must be considered because images naturally suggest dream and fantasy. The question then becomes, what wishes and fantasies do viewers bring to media environments?

THE WISH

Using a psychoanalytic interpretation of viewing film and television, I suggest that those films that are most successful tap into personal fantasy structures. If this is true of the film medium, we need to then consider how this use of fantasy can help students learn. However, what determines whether or not we like a film is more complicated than simply if we would "enjoy" experiencing each particular film fantasy—there are other issues at work. American films are about power—gaining it and losing it. Howard Suber of the School of Theater, Film and Television at UCLA claims that one can describe all American comedies as being about a character who lacks power at the beginning and who then attains it in the end. Conversely, all American dramatic films are about a character who begins with power and then loses it (*Citizen Kane, Wall Street*). In this formula of power exchange in American film, the commercial requirement of the need for happy endings tends to modify this general structure. Thus, many dramas have the main character lose something during the main action, but then gain something else—often more important—in the end. In *Rain Man*, Tom Cruise loses his father's inheritance but gains a brother. Similarly, comedies often end with the main character losing what was sought, while gaining something else. Although the overall movement of the action in *City Lights* is towards Charlie gaining power and the gratitude of the flower girl, he loses her illusion of him as a rich businessman. In both cases, the formula of power exchange holds true as an overall process in the film, if not in the specific type of ending. Conflict in both comedy and drama is always a struggle for some sort of power, whether it is dominance in a personal relationship, in a family, in business, or in the political arena.

It is useful to look at the more subtle ways in which the exchange of power is utilized in film from themes, to dialogue, and even the manner in which scenes are staged. Directors sometimes talk about the need to understand the exchange of power between characters in a scene; the character with power should control the camera and the staging. In *Touch of Evil*, look at the way Orson Welles' staging emphasizes the struggle between the characters by following the dominant character with the camera. This technique is used repeatedly, consciously or unconsciously, in many American films.

Deeply rooted in the American psyche is the myth of social mobility and the desire to climb the ladder in our capitalist economic system. It is hardly surprising that a popular art form that developed during the great expansion of American society—one that was readily accessible to all classes, even those who could not yet speak English—would pay to see these collective American dreams of ambition. The only thing that changes in this popular film formula of ambition is the object of desire, the symbol of power. How can this basic formula of American drama be used in developing computer-based educational environments? Education is a kind of power. One of the motivations for gaining education is because it is connected to income, social position, and worldliness. Perhaps the pursuit of power seen in the story structure so prevalent in American films can be used in the design of computer applications.

THE HERO

The primary vehicle for film fantasies is the hero. For Freud, the way that Western literature and film are concerned so completely with the hero is a result of the egocentric nature of dreams: "His Majesty the Ego, the hero of all day dreams and all novels" (Freud, 1963, p. 40). It is through the hero that the audience enters the film world. This process is one of identification; each member of the audience sees not himself, but the idealized self in the hero. It is important to note here that stories don't have to be told in this manner. Many Japanese movies (such as Yasujiro Ozu's) have an expanded notion of subject, reflecting a non-Western world view, one that is not egocentric and preoccupied with a main character.

To say that the audience of a Hollywood-style film dreams of being the hero of films—to be good looking, worldly, and powerful—is commonplace. But what is the specific mechanism of this fantasy structure? What techniques does a writer use to manipulate this tendency of the audience to identify with the hero? In terms of premise and the types of stories that Hollywood filmmakers tend to tell, the requirement of a main character who appeals to the audience is paramount. One simply does not see very many main characters that are poor, weak, physically unattractive, or powerless. The last characteristic, power, is especially important. Since American films are often about the exchange of power, a premise must incorporate a pursuit of power in the main character's motivation. For educational media products, this strong principle of identification has political and ethical ramifications. Should this identification principle be utilized? If so, are only the traditionally powerful and attractive to be used as heroes?

BUNUEL

In looking at the literature on the psychoanalytic approach to filmmaking, one would be remiss not to discuss the work of the great Surrealist filmmaker Luis Bunuel. What is useful to consider with Bunuel is how his intentions to confront the viewer might have application in computer environments. The use of dream sequences in Bunuel's films comprise one of the most distinctive elements of his narrative style, often providing a core of meaning that resonates throughout each work. In the context of each film, these dream images constantly suggest more than themselves by their parallel situating: the dream and the story comment upon each other and suggest a meaning beyond each individual expression.

Bunuel's film language greatly reflects the various structuralist movements in philosophy; he is concerned with developing a new sort of meaning structure. Martin Heidegger, the important 20th century philosopher, concluded in his later years that it was the distinct role of great art to create new authentic patterns of meaning. Heidegger (1962) discussed the importance of metaphors noting that in authentic art, a bringing together of usual signs and signifiers in such a way that they evoke a combined meaning beyond the separate terms. This is quite similar to the Surrealist practice of making so-called "exquisite corpses," collage type drawings created by different people and assembled to vaguely represent the human body (Figure 38). This Surrealist approach is the reason Bunuel films are often hard to categorize in traditional terms.

It is useful here to look at how Bunuel implements his surrealist strategies. Generally, his films make people feel uncomfortable, partially because of his violent tone, but mostly because of a narrative style that constantly reaches beyond itself. In many of his films, Bunuel tries to disrupt the viewer from looking at the work in a purely linear rational manner. In the opening of *Un Chien Andalou* when an eye is cut by the razor, the viewer's comfortable rationalism is attacked almost immediately. Similarly in *Belle de Jour*, when the film begins in a daydream sequence, the main character is awakened by asking: "What are you thinking?" In both instances, Bunuel forces the audience to look at each sequence in relation to the film as a whole: he forces the viewer to make associations that call forth a rich meaning.

Figure 38: Exquisite Corpse—Two Ambiguous Figures (Max Ernst, 1919)

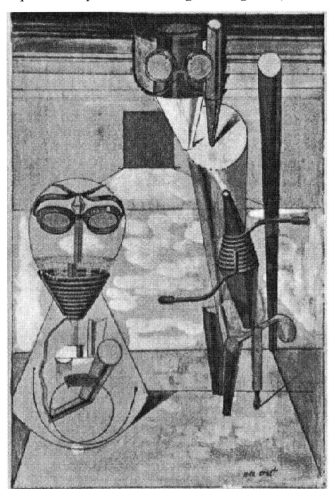

In *Los Olvidados*, a film made in Mexico, Bunuel takes a seemingly straight forward social condition story and dream and makes it a rich, complex work. He constantly combines images of motherhood and violence. For example, the central dream first shows Pedro being comforted by his mother in bed. In slow motion, she asks him why he helped kill another boy and tells him to look at her hands rough from scrubbing. This sequence expresses at once motherly love and violence. In this central dream sequence, the motherly love expressed at the beginning, with the medium-shot of Pedro embracing his mother, is combined with the grotesque image of his mother offering a piece of raw meat to him and having it taken by a dead boy from under the bed. Thus Bunuel combines imagery to depict the film's central themes: motherhood, violence and social depravity.

In this central dream sequence Pedro asks his mother for something to eat, and she gives him a piece of raw meat, much like a mother bird feeding its chick. Birds are used continually throughout the movie to form a web of meaning and seem to serve much the same function and possess the same connotations as Pedro's mother. Taken as a whole, the bird images seem to represent motherhood gone wrong. Bunuel visually emphasizes the connection between motherhood and birds by having feathers fall across the medium-shot of Pedro's dream about his mother. After the gang of boys beat up the blind man, Bunuel pans from a close-up of the blind man on the ground to a chicken. He then cuts to a close-up of Pedro playing tenderly with a hen who has laid an egg. Here then, Bunuel visually connects an image of motherhood gone wrong in creating children who beat up blind men, with a shot of innocent and natural motherhood. Conversely, when Pedro is angry he vents his anger against birds by killing the chickens in the boy's home and, in this way, Bunuel illustrates how Pedro has lived an ill-formed childhood.

In many ways, *Los Olvidados*' character portrayal is probably the most sympathetic of all of Bunuel's films. Bunuel uses a first-person camera position at the beginning of the film to represent a boy acting as a bull running at a coat. Making the audience assume the position of one of the boys is a kind of sympathetic approach to the development of that character. This film also has a musical sound track, unlike most of Bunuel's films, that reflects the character's predicament or emotional state. Bunuel uses a weeping flute sound when showing Pedro walking around the city after running away from home. Bunuel uses his original Surrealist cutting style to combine various images of motherhood, birds, and violence. Throughout, he cuts images of love with scenes of violence and cruelty. Out of what could have been a straightforward story about slum life, Bunuel has constructed an unsettling, Surrealistic roller coaster of dream-like images.

In *Los Olvidados*, Bunuel has created a narrative that clearly revolves around a cluster of dream images. The sequence of events in the film constantly relate back to these dream images. Each event provides a new context for the images and thereby creates a different kind of meaning. The meaning structure is expanded as the storyline continues to evoke further connections. The images are not simply symbolic: birds don't only represent motherhood. By putting the various images in different and evolving contexts, Bunuel enriches their meaning. Violence and motherhood are not normally images that are related; it is a distinctly Surrealist pre-occupation to combine seemingly unrelated things in order to call forth a new meaning.

Belle de Jour begins with a daydream. As in *Un Chien Andalou*, Bunuel again trys to disrupt and broaden the viewer's interpretive faculties by forcing him or her to give equal weight to all the images, whether they seem to be dreams or not. The first sequence contains the film's central masochistic contradiction: an unhappy event combined with Severine's pleasant reaction. While the viewer sympathizes with Severine over the male brutalism she suffers, Bunuel's cut to a low-angle close-up of Severine's face, which reveals her partly pleased expression, repulses the audience. The film's various fantasies and dreams form a narrative structure that expresses the complexity of Severine's pleasure/pain confusion.

In addition, Bunuel creates several childhood memory scenes that offer a kind of psychoanalytic explanation for Severine's masochism. Bunuel inter-cuts a series of close-ups on Severine as a child refusing to take the communion bread in her mouth, with her first meeting with Madame Anais who introduces her to prostitution. Bunuel connects memory with present time by zooming in to a close-up on Severine while a voice chanting in Latin is heard. Another scene of memory for Severine is when she recalls a plumber kissing her on the cheek just as her mother calls for her. Her desires and confusion of childhood have remained part of her present because she is unable to understand herself.

The various dream sequences in *Belle de Jour* all illustrate the masochism of Severine. Being raped, having mud thrown at her, and imagining her husband paralyzed characterize Severine's self-lacerating quality. The viewer sees most of the film from her viewpoint and therefore must rely on her fantasies and memories for narrative development. The viewer is made to take part in Severine's masochistic imaginings because there is no narrative outside of her. This is why people often feel uncomfortable watching this film, and also why the ending is difficult to interpret. By relying on a masochistic character's mental or psychological internal state to tell the story, Bunuel has fashioned a distinct narrative style.

The use of dream-like images for expression is particularly well suited to the medium of film, with its great ability to imitate life, its ability to re-order time through editing, its ability to represent internal perception of an individual through visual narration, and the likeness of watching a film in a darkened room with the sleeping state. Bunuel's narrative style is distinctive because it depends greatly upon a cluster of dream images. Most of his films are totally unintelligible if understood in a strict rational manner. Bunuel, as a member of the avant-garde, searched for deeper structures of meanings through new artistic methods of expression. Bunuel is a great artist because he successfully expresses an original meaning through using his medium to its full potential. It is interesting here to compare the way Bunuel uses visual metaphor in shifting patterns with the way Eisenstein uses a physically based metaphor in a dialectical formation. Bunuel relies more on the viewer to construct personal meaning from the association and consequently ends up creating more complex narratives. For computer learning environment designers, Bunuel demonstrates a very sophisticated approach to the construction of narrative and an artistic sensibility that could be useful as a model for inciting deeper thinking and imaginative thoughts in learners.

CONCLUSION

Film as a language has developed to the point that viewers understand the importance of context and order, or the syntax of images. The film viewer assumes intentionality in the order of images, but this may not be the case with the computer medium because of user control in navigation. In some ways the editing principles of film work well in computer environments, in other ways they are more problematic. As an older medium, film provides ready-made conventions somewhat automatically employed by viewers when using computer software. While this tendency brings advantages in easily communicating with the learner, it also lends itself to passive viewing and ideological slanting of content. Psychologically, "high concept" allows the audience members to feel the safety and satisfaction of seeing a "new" story told, which leads in a direction they already know, because they've seen the same story time and time again. The story will not turn in any perilous, uncharted direction or, worse yet, end in a disturbing manner. So, unlike everyday life, the audience gets the immense existential satisfaction of knowing beforehand the fate of a fictional character in a fictional world. In this way, Hollywood-style films utilize fantasy structures in a very narrow fashion—principally, only in such a way that will make money at the box office. Designers of educational software need to be aware of and understand this psychological principle and use it to advantage.

In the next chapter, we turn to the growing field of narrative psychology and the specific techniques of stories, simulations, and case studies used in computer learning environments.

Chapter XII

Stories, Simulations, and Case Studies

Recent literature in the field of psychology and education connects storytelling and narrative construction with learning. Some even see storytelling as the foundation of teaching (Abrahamson, 1998). In this model, teaching is seen as storytelling, and learning becomes personal story construction. In this chapter, I examine the argument that narratives frame knowledge and experience and are central to the process of teaching and learning. I then look at how the use of narrative as a structuring device works in educational applications and through the specific techniques of computer simulations and case studies.

In the 2001 study, more than half of the respondents indicated story formats were helpful in the presentation of course content (Figure 39). In response to the statement, "When course content is put into the form of a story it is easier to understand," 66.0% either strongly agreed or agreed.

The 2001 study also found that students preferred the use of simulations and case studies (Figure 40). 85.4% of the students either strongly agreed or agreed with the statement, "The use of case studies and computer simulations increases learning for me."

Figure 39: Course Content in Story Form Easier to Understand (Questions 1 & 16)

Delivery format * When course content is put into the form of a story it is easier to understand. Crosstabulation

			When course content is put into the form of a story it is easier to understand.				Total
			strongly agree	agree	disagree	strongly disagree	
Delivery format	computer-based	Count	7	17	17	4	45
		% within Delivery format	15.6%	37.8%	37.8%	8.9%	100.0%
	videotape	Count	4	32	12		48
		% within Delivery format	8.3%	66.7%	25.0%		100.0%
	correspondence	Count		3			3
		% within Delivery format		100.0%			100.0%
	other	Count	1				1
		% within Delivery format	100.0%				100.0%
Total		Count	12	52	29	4	97
		% within Delivery format	12.4%	53.6%	29.9%	4.1%	100.0%

Figure 40: Case Studies and Simulations Increase Learning (Questions 1 & 21)

Delivery format * The use of case studies and computer simulations increases learning for me. Crosstabulation

			The use of case studies and computer simulations increases learning for me.				Total
			strongly agree	agree	disagree	strongly disagree	
Delivery format	computer-based	Count	12	28	6	1	47
		% within Delivery format	25.5%	59.6%	12.8%	2.1%	100.0%
	videotape	Count		1			1
		% within Delivery format		100.0%			100.0%
Total		Count	12	29	6	1	48
		% within Delivery format	25.0%	60.4%	12.5%	2.1%	100.0%

It is likely that the difference in responses to the two questions has to do with some of the respondents not understanding how narrative might be used in computer contexts. When specific uses of narrative through case studies and simulations is mentioned the rate of agreement rose almost twenty percent. Similarly, administrators overwhelmingly agree when the same question is posed (Figure 41). The 2000 study found that most institutions surveyed (88.5%) use simulations and case studies.

Figure 41: Courses Include Simulations and/or Case Studies (Question 43)

Courses include simulations and/or case studies.

		Frequency	Percent	Valid Percent	Cumulative Percent
Valid	strongly agree	36	20.5	21.8	21.8
	agree	110	62.5	66.7	88.5
	disagree	16	9.1	9.7	98.2
	strongly disagree	3	1.7	1.8	100.0
	Total	165	93.8	100.0	
Missing	System	11	6.3		
Total		176	100.0		

However, simulations and case studies were used far less than any other of the specific pedagogical approaches—undoubtedly, partly a result of the expense and difficulty in their creation. In interviews, some administrators noted this method:

> We have one teacher who does contract negotiation courses. He does teach simulations to teach the concepts and skills of negotiation. We put the simulations up on the Web with directions for the student who is the buyer and the student who is the seller, and it works (Warren Ashley, Director of Distance Learning, CSU Dominguez Hills).

> For the meteorology course, students have to go onto the Web to pull down data. We're offering a course in personal finance in which students do their own plan and probably do a simulation (Arthur Friedman, Professor and Coordinator, College of the Air, Nassau Community College).

> Yes, I'm not sure of the software used, but I know in some courses they are used. Something as simple as Excel is used, and they do some CAD work as well (Elizabeth Spencer-Dawes, Manager, Distance Learning, Boston University).

Many more respondents reported using case studies. However, rather than identifying them as a pedagogical technique well suited to distance learning, most simply carry over typical practices from traditional courses to distance learning formats.

> Well, case studies have always been part of what I do, especially for special ed[ucation]. We use a lot of case studies in special education, but in research methods, the case studies tend to be vignettes describing a study and then figuring out where they went wrong. What could they have done differently? How could they have controlled better for internal validity? That type of thing. That is part of their online learning (Don Cardinal, Chapman University, School of Education Faculty member).

As the following respondents note, case studies are often used in business school programs.

> We usually have several School of Business courses on interactive television, and they rely heavily on case studies (John Burgeson, Dean, Center for Continuing Studies, St. Cloud State University).

> The operation management courses tend to be case based so there's a lot of interaction in the classes (Elizabeth Spencer-Dawes, Manager, Distance Learning, Boston University).

The following respondent notes a common occurrence in business courses tending to use case studies, while the sciences often use simulations.

> Yes, I'm thinking of one of our business courses, he [the faculty member] uses case studies a lot. Our astronomy teacher uses simulations that he's built (Greg Chamberlain, Dean of Learning Resources, Bakersfield College).

In addition to business, law is an area that also uses case studies extensively.

> I think some of the business folks do. I think the business law instructor used them (Thornton Perry, Director of Distance Education, Bellevue Community College).

As with other pedagogical approaches, some institutions report faculty members determine the use of simulations and case studies.

> Yes, whatever the faculty comes up with, almost anything you could think of (Carole Hayes, Coordinator, External Relations and Development, Office for Distributed and Distance Learning, Florida State University).

> As they are in any [institution] courses (Program Manager, anonymous large, independent, Western U.S. doctoral degree-granting institution).

Others report not using simulations at all.

> We do not (Joy Edwards, Director of Graduate Studies, Texas Wesleyan University).

> Not that I'm aware of (Jacquelynn Sharpe, Division of Distance and Distributed Learning, Georgia State University).

Because of the expense and labor involved in creating computer simulations, one is less likely to find them in common practice. Case studies are often text-based and already part of the current curriculum, therefore easily incorporated in distance learning format courses. In summary, the data on current practices show that both students and administrators recognize that case studies and simulations can be effective in learning environments. However, simulations are used sparingly and the more abstract notion of the role narrative might play is less understood. Although case studies and simulations are types of narratives because they use a story structure, most students and administrators surveyed did not appear to have this broader understanding.

STORIES TO ORGANIZE LEARNING ENVIRONMENTS

In the field of computer interface design for educational settings, interest in storytelling and narrative construction structures learning is growing. Because computers allow users to quickly and easily manipulate and build personal stories, they potentially offer learners a powerful tool. Brenda Laurel's work (1990, 1993) in particular explores how computer-based education might be presented as theater. At the British Open University, a research group called "Meno" (Laurillard, 1997; Plowman, 1994, 1996, 1997) has looked at storytelling and narrative in the construction of educational computer-based programs.

In viewing computer interface as a medium rather than a tool, some (Laurillard, 1997) have noted a need for a new symbolic system or language for computer interfaces. As a developed medium with many parallels to computer interfaces, contemporary film theory is relevant. As we saw earlier, writers in the cognitive movement in film theory (Bordwell, 1989; Branigan, 1984; Carroll, 1988) see the process of viewing a film as a meaning-making process. This theory and others hold important implications for interface design in educational settings.

The key issue arising through the examination of the diverse literature is: how are narratives in interactive hypermedia environments constructed to make learning most productive?

REVIEW OF PSYCHOLOGY AND EDUCATION LITERATURE

Probably the most influential figure in the promotion of narrative in cognitive theory is the psychologist Jerome Bruner. As opposed to other cognitive theories that are information database models, Bruner's theory is based on meaning making. He argues that meaning making comprises psychology's central pursuit and is the essence of the so-called "Cognitive Revolution." In summarizing the Cognitive Revolution, Harre and Gillett (1994) broadly describe the view of the mind as a social construction controlled by discourse. According to Bruner (1986), the Cognitive Revolution developed in the late 1950s and focused on how individuals acquire and use knowledge, signaling a philosophical shift from a behavioral emphasis on performance to the centering on competence. The development of narrative psychology grew out of the Cognitive Revolution, with psychologists borrowing from fields such as literary criticism, philosophy, and anthropology to establish a new culturally based approach. Focusing on how humans respond to experience by constructing stories, this theory posits that stories are the tools by which cultural meaning is constructed and communicated.

Bruner (1990) proposes that it is important to discover how meaning making occurs in a larger cultural arena, and that narrative functions as one of the primary tools for its creation. Individuals construct versions of themselves in the world through narrative; similarly, it supplies whole cultures with models of identity. This narrative approach has also been recently used in clinical environments (McLeod, 1996; White & Epston, 1990), where therapists interpret their clients lives through the telling of the latter's personal stories, and then intervene with the reconstruction of these stories.

Bruner (1986) was one of the first cognitive psychologists to articulate the distinction between the narrative and the paradigmatic (or abstract) dimensions of knowing. In this dichotomy, stories and arguments embody the two kinds of meaning making. Arguments persuade one of truth, while stories convince the learner of "lifelikeness." Deslauriers (1992) claims that narrative is context-rich, while the paradigmatic aims at being context-free. In examining the notion of meaning making, Bruner argues that meaning cannot be understood when removed from a cultural context. This is one of the central advantages of using stories in education—it creates context-rich knowledge.

Figure 42: Narrative Versus Abstract Knowledge

Stories/Narrative	Abstract/Paradigmatic
context	objective
culture	non-specific
lifelikeness	removed
action	behavior
competence	performance
personal	anonymous
1st person	3rd person
active integration	passive reading
retelling	dominant stories
cases	rules

Bruner promotes what he calls "folk psychology," an understanding of mental processes that incorporates culturally derived meaning with personal experience. Folk psychology is a culture's account of what makes us human. This kind of cultural psychology emphasizes action over behavior, or action situated in a culture. Bruner argues that the search for meaning in culture should be central to human endeavor. Folk psychology is a system whereby individuals order their life experiences, with narrative as the organizing principle. For Bruner (1990), the cultural setting of one's actions forces children to become narrators. He asserts that children quickly learn that getting what you want in the world means creating the right story. Finally, Bruner believes in the constructivist tenet of reality being constructed, not found. Thus, education is assisting in finding the tools of meaning making and reality construction. He argues that narrative skills need to be taught and points to the need for metacognition, where the learner is made aware of his own thought process (Bruner, 1996).

In psychological practice, the new field of narrative therapy has emerged over the past few years. The Australian psychologists, White and Epston (1990), look at clinical practice and find that human beings are interpreting beings and that it is not possible for humans to understand experience without access to some structure of intelligibility, or meaning giving system. They argue that stories are just this kind of framework of intelligibility. Proposing that stories in a clinical setting can shape lives through active interpretation, they see the use of alternative story models as a good clinical approach. White and Epston use Bruner's theories to suggest that stories have dual landscapes of action and consciousness. Their clinical "re-authoring" work leads patients to reflect on alternative events in their personal landscapes, and to consider motives and characteristics of the story. In their influential book, *Narrative Means to Therapeutic Ends* (1990), White and Epston focus on externalizing the clinical problem through storytelling, primarily in the use of letter writing. "Externalizing" is one of the key concepts in their approach that encourages individuals to objectify problems and make them external. They draw heavily from Foucault, the French philosopher, in analyzing the social context of therapy and the dominant stories that structure the clinical experience. Individuals in the clinical setting have problems when these dominant stories do not match their individual reality. Thus, dominant stories need to be challenged and retold, according to White and Epston. This research raises a critical issue for educators: one needs to think about the ramifications of telling and interacting with dominant stories in educational environments.

McLeod (1996) argues that the fundamental philosophical position of a narrative approach to counseling is constructivist. In this viewpoint, the self is constructed by drawing on common cultural stories that provide a way for individuals to accommodate cognitive structures. Because narratives are structured over time, they emphasize order and sequence, which is useful in a clinical setting. Additionally, McLeod proposes that narratives should use an active protagonist with whom the self can identify.

Abrahamson (1998) argues that storytelling forms the foundation of teaching because learning is essentially the understanding of new events in the context of past events. In reviewing the literature, he claims that research has shown putting educational content into a narrative format helps students to think critically and to personalize their learning experiences. Storytelling is an effective approach to education because it is motivational, personal, can incorporate cultural and ethnic identity into the learning process, and helps to create cognitive pathways.

Most important, Abrahamson argues that storytelling directly inserts the teacher's—the course storyteller's—point of view. Rather than presenting course content objectively, storytelling allows the teacher to insert consciously his/her own perspective. This approach allows the student to better position herself/himself within the teacher's cognitive structure, while at the same time understanding the learning experience is not objective. This point is useful to consider when looking at the point of view issue raised earlier. Storytelling also works well as a vehicle for cultural expression. In this model, the curriculum becomes a collection of cultural stories, with teachers the tellers of cultural tales (Abrahamson, 1998).

Some theorists characterize the story listener's psychological state as being mildly hypnotized. Erikson, Rossi and Rossi (1976) argue that hypnosis involves taking a subject's unconscious knowledge and revealing it to him or her in hypnotic states. When students listen to a story, they often suspend their usual cognitive framework and learn in a different manner. In fact, Erickson, Rossi, and Rossi define the hypnotic trance as "unconscious learning." They propose that students are most open to learning in this state because usual frames of reference are temporarily disabled, and they therefore receive new ideas more freely. According to Erickson, Rossi, and Rossi, the most effective way to induce the learner's trance is by acknowledging student life experience. Students can then reorganize new ideas and accommodate them into their own cognitive structures.

This brief summary of the research literature reveals that strong connections exist between storytelling and learning. Important practical questions need to be addressed regarding how storytelling is used in computer-based learning environments. One particularly fruitful use of storytelling in computer software design is to lessen the problem of learners getting lost when following hypertext threads. The British Open University's Meno Project is charged with studying narrative in computer-based educational programs for this specific reason. Diana Laurillard (1997) describes the essence of Meno's paradox as, "How can you learn something you know nothing about?" For Laurillard, the question revolves around how learners manage the process of understanding and meaning making, while sorting through endless webs of connected knowledge on the Internet or through computer databases. Is there a technique or approach that can be used? Laurillard reports that the presence or absence of narrative structure in multimedia programs directly affects the learner's comprehension level. She points out that many identify the ability to have multiple choices and variables in computer-based storytelling as an advantage, but that in learning situations this can lead to confusion and lack of focus. Consequently, while interactivity has been viewed as an advantage, it may in fact be a disadvantage when used for educational purposes.

Johnson (1997) traces the rise of the 19th century novel to the function of explicating new social realities, while 20th century computer interface addresses the overwhelming amount of information in search of meaning. Thus narrative in computer environments becomes a mix of metaphor and footnote and is consequently atomized. Johnson concludes that graphic user interfaces should be organized around meaning instead of space—in direct contrast to the reality, where most GUIs are organized around space. Consequently, the research literature refers repeatedly to the problem with users getting lost in hypertext space.

NARRATIVE STRUCTURE IN HYPERTEXT ENVIRONMENTS

A review of the literature in the fields of psychology, education, literary theory, computer interface design, and film theory reveals a growing interest in narrative as a fundamental tool for the creation and expression of culturally derived meaning. In terms of narrative psychology's practical applications to computer learning environments, much remains to be done and many basic issues need to be resolved. The following primary questions emerge.

First, what is the shape of narrative structure in hypertext environment with multiple threads? Bruner argues that the narrative's characteristics are that established traditions (or canons) of storytelling supply a sequence to meaning making and focus on individual characters. Additionally, narratives serve to transmit cultural beliefs and norms, while also allowing for some variation among individuals. This tension between cultural and individual meaning is one of narrative's key aspects. Narratives include "negotiated meanings"—canonical social norms and expectations, as well as exceptions to the rule. Indeed for Bruner, the function of the story is to find something that makes understandable a deviation from the traditional cultural pattern. How is this done?

Reason and Forrester (1997) argue that in order to create narrative using hypermedia, a determined ending is necessary. They argue that soap operas with an episodic narrative structure provide a model for the use of narrative in hypermedia because of their open-endedness. They claim that hypermedia stories still require established endings tied to the preceding narrative, but do not need to have a firm sequential order. Plowman (1996) argues that the author is displaced in interactive multimedia. However, this notion of the user as author is often an illusion because the user only chooses between options created by the designer, and cannot create anything original. Perhaps nonlinear approaches can be used within traditional narrative forms. If so, how is this done?

This suggests both a new model for content and the need for a narrative approach to browsing which incorporates some partnership between the viewer and the instructional designer in authorship. Maybe the audience needs to be able to steer the presentation dynamically so that it relates to their interests. Murray (1997) proposes that "memory frames" allow for multiple particularization of story elements without destroying the story. For her, the creation of a commentary space in a digital world would create a kind of Greek chorus to comment on the action.

Seymour Chatman (1978) argues that narrative functions through transformation and self-regulation. Self-regulation involves Piaget's notion of structures never leading beyond the system, but always engaging elements that belong to it and preserve its laws. Successful narratives will not admit scenes or elements that do not belong to them. This would seem to put a real limitation on learning. Again, how can endings be predetermined while the sequence is flexible?

Looking at the use of narrative in hypertext educational environments, the possibilities are matched only by the questions about implementation. One way of beginning to understand implementation is to look at the way that two of the early forms of narrative—case studies and simulations—are currently used.

CASE-BASED LEARNING/SIMULATIONS

According to Jonassen (2000), case-based reasoning (CBR) is employed to analyze stories and experiences for what they have to teach us. Simulation is a major part of the research literature on HCI, particularly as it applies to educational environments. An early review of the literature showed the number of published simulation articles at approximately 200 for each of the years 1986 to 1990 (Pickover, 1991). In the last decade, the research literature has grown further. Especially in educational environments, simulation can be a very effective tool. Because of the rich multimedia computer environment, learners can better bridge the gap between reality and the simulated task. Learning by doing is accomplished through simulation and is especially useful where actual environments are expensive and impractical to recreate constantly (Feifer, 1994). Simulation is effective because it can create a context for learning. Pickover encourages the use of the computer as an instrument for both simulation and discovery, particularly in science.

Feifer argues that the difficulty of creating good simulations and of learners using them by themselves are two factors limiting their use in teaching.

Roger Schank (1997), a leader in the field of designing computer learning environments, argues that learning in computer simulations, or virtual learning, offers the best opportunity for students to learn by doing in an apprenticeship-type model. He states that one of the biggest issues for learners is that they have trouble failing in public, while computers offer an ability to fail independently. Schank focuses on the use of stories in simulators and on both expert and non-expert storytelling in simulations. He argues that in workplace learning, stories form the root of organizational knowledge. By simulating scenarios based on common organizational stories, employees can quickly acquire needed knowledge.

Among others, Schank (1997, 1998) began by looking at artificial intelligence models that led him to investigate the use of narrative in cognitive structure. Early on, Schank argued that memory is organized around personal experiences or episodes, rather than abstract semantic categories. He defined understanding as the process by which people match new experiences with stored impressions of past experiences. Schank asserts that understanding is the process of index extraction and the proper juxtaposition with related indices. He argues that we primarily learn from a re-examination of our own stories. Schank points to insight as the recognition that old stories have new relevance and use. "Gist" is Schank's notion of how the mind remembers a particular story's significant structure. Schank holds that we assess other people's intelligence based on their ability to tell stories and to listen to ours. Thinking depends on storytelling. Intelligence is not problem solving, but the ability to address situations without answers. He argues that "scripts" are story segments used in conversations to answer questions and respond. They provide set, expected answers. The more scripts one knows, the more one is able to move comfortably in different social worlds.

Schank (1997, 1998) indicates that the use of storytelling in education can work well in representing real life experiences. He argues that rather than teaching rules, we should teach cases (stories) and the adaptation of cases. Case studies have long been a teaching method in fields such as business, medicine, and law. Much of what students in these professions learn is how to apply past cases to new situations. Apprenticeship models are based on the notion of a kind of one-on-one storytelling. With the advent of computers, simulations can also provide this kind of reality-based story work with computer-generated scenarios. The reason this kind of realism can be provided by the narrative structuring of educational content is because it is context-rich. Jerome Bruner's notion of lifelikeness is a result of this richness. Furthermore, an important aspect of this context richness is in providing cultural models of identity (Bruner, 1990), which are very useful in serving diverse populations.

CONCLUSION

This chapter reviewed the growing research literature in the field of narrative psychology and examined the role that narrative plays in human thinking and the creation of meaning. The implications of narrative psychology in the design of computer-based educational environments has been recognized by such leaders in the field as the British Open University and the American researcher Roger Schank. Clearly, this is a rich area for further research. In the next chapter, I summarize what we have learned from the theories and data presented in this book.

Chapter XIII

Conclusion

medium, n: 1. A middle quality or degree of intensiveness, etc. 2. A substance or surroundings in which something exists, moves, or is transmitted; an environment (Hawkins, 1986, p. 521).

tool, n: 1. A thing (usually something held in the hand) for working upon something. 2. A simple machine, e.g., a lathe (Hawkins, 1986, p. 867).

A computer is not only a simple machine, but a significant new surrounding in which learning exists, moves, and is transmitted. We began this book by asking key questions about how computer-based educational environments are created. What are the specific properties of the new medium? What are its specific advantages for education? What content is best delivered through this medium? What are the most effective techniques for the design of computer learning environments? We examined the notion of computer as medium and what such a notion might mean for education. Finally, I suggested that the understanding of computers as a medium may be a key to re-envisioning educational technology. In this chapter, I pull together the data presented from the two studies, filter it through the research literature discussed from a wide-ranging and eclectic group of fields, and draw some conclusions.

In order to address the questions posed at the beginning of this book, I divided it into two parts: the first part focused on computer-based learning and theory and practice, while the second part looked at the issues connected with considering the computer as a medium. In other terms, the first part was about the things we know from years of research, while the second part is about a vision for the future arrived at in this book through combining science and the humanities. In Chapter Two, a brief analysis of the history of both educational technology and relevant learning theory was presented. In the next two chapters, the two main approaches (tutorial and group) to distance learning were examined. In Chapter Three, the specific methods and issues connected to the tutorial methods were discussed with a particular emphasis on customization, profiling, and agents. In the fourth chapter, group learning methods were investigated. The research literature on virtual teams, interdependent tasks, community formation, and specific techniques for organizing group tasks in computer educational environments was discussed. Then in Chapter Five, the literature on human-computer interaction was examined with an eye towards understanding how it may be especially relevant to education. In the following chapter, the issues of interactivity and navigation in educational environments were then examined. In Chapter Seven, computer tools for learning such as concept maps, knowledge modeling and learning through computer programming were considered. The first part of the book ended with a look at teaching and faculty issues.

In the second part of the book, we considered the ramifications of understanding computer software used for educational purposes as a new medium. In Chapter Nine, broad issues such as phenomenology and ideology of film were examined, as were the history of film and still photography, documentary film and recent film theory. In the following chapter, dramatic structure, genre and editing were considered. In Chapter Eleven we looked at specific issues such as point of view, subjectivity, and the psychoanalytic interpretation of the process of viewing. In Chapter Twelve, the ramifications of the recent narrative psychology movement on computer environments were examined particularly as they relate to the use of case studies and simulations.

The conclusion of this book is divided into three sections. The first section describes the current practices of distance learning and approaches to designing educational software. The second part discusses the conclusions of the research presented that, to my mind, clearly indicate certain guidelines for good practice in educational software design. The third section explores visionary notions of what distance learning in the future might be. I've structured my concluding remarks thus because it is important to separate out, in this early stage of development of distance learning, what we know with some degree of certainty from what we imagine or hope for in the future. Oddly enough, the scholarly work on distance learning has been hampered by both a lack of concrete knowledge and vision. By vision, I do not mean the exaggerated claims of traditional education's downfall or the mechanization of learning; rather, how distance learning in the future might lead to more pervasive and integrated notions of learning for adults.

Figure 43: What We Know/How We Might Learn

Needs Current Practices	What We Know Best Practices	How We Might Learn Future
Media:<ul><li>not designed for medium</li><li>low media sophistication</li><li>licensed software</li></ul>Population:<ul><li>growing</li><li>diverse</li><li>voluntary</li><li>older</li><li>part-time</li><li>employed</li></ul>Methods:<ul><li>behavioral</li><li>constructive</li></ul>	Methods:<ul><li>reexamine traditional teaching/learning practices</li><li>use learning theory appropriate to objectives</li><li>beware ideological transmission</li></ul>Techniques:<ul><li>communication with faculty a key</li><li>increase interactivity with other students</li><li>use effective navigation schemes</li><li>borrow from older media</li><li>emphasize tutorial and group learning methods</li><li>learning through programming</li><li>use concept maps</li><li>use common computer applications for learning</li><li>use simulations</li><li>make learners aware of point of view</li><li>limit voice-over narration</li><li>make learner aware of meaning-making activity</li><li>create communities</li><li>focus on learner usability</li></ul>	Methods:<ul><li>increase informal learning</li><li>integrate learning with work</li><li>use narrative</li><li>use dreams</li><li>use artistic methods to challenge learners</li><li>accommodate learning styles</li></ul>Techniques:<ul><li>visualization</li><li>intelligent tutors</li><li>let users create personal stories</li></ul>Learning Devices:<ul><li>single-use learning devices</li><li>facilitate communication</li><li>facilitate interdependent tasks</li></ul>

CURRENT PRACTICES

The 2000 survey results show that most higher education institutions have unsophisticated approaches to developing distance learning courses (Figure 44). Only a small percentage (28.1%) of 2000 administrative survey respondents indicated that they develop distance learning format courses specifically for the platform.

Figure 44: Courses Developed Specifically for Format (Question 48)

Which of the following most closely reflects the course development process?

		Frequency	Percent	Valid Percent	Cumulative Percent
Valid	existing course material automated by technology	26	14.8	15.6	15.6
	existing course material automated, with some new material	94	53.4	56.3	71.9
	all course material developed specifically for DL courses	47	26.7	28.1	100.0
	Total	167	94.9	100.0	
Missing	System	9	5.1		
Total		176	100.0		

For the most part, educational institutions are simply taking traditional classroom methods and automating them with computers and other media. The reasons are many, but the approach is clearly wrong.

A review of the literature about higher education reveals the following important and relevant trends: the need for higher education is increasing greatly; the student population's composition and characteristics is changing drastically; learner-centered approaches are becoming more widespread, particularly in computer-assisted education; and the need for the integration of education and work is increasingly recognized. First, demographic forces and economics, as well as political and lifestyle trends all show a growing need for non-traditional education exists (Heterick, 1993). While particularly dire outside the United States (Daniel, 1997), even California faces an estimated increase of 500,000 students in higher education institutions in the next few years (CPEC, 1997). The characteristics of distance learning students fit this changing demographic group: they tend to be voluntary, motivated, have high expectations, possess more self-discipline, and are generally older (Palloff & Pratt, 1999). The single most important factor is the changing demographics of higher education, which leads to a great increase in the student's average age. One-third of all undergraduate-level and two-thirds of master's-level enrollments are part-time. Women comprise the largest single demographic group among part-time degree students. In increasing numbers, adult learners will have the following characteristics: computer skilled, educated, employed in work that is primarily mental (Peter Drucker's knowledge workers), and accustomed to cooperative work (Davis and Botkin, 1994; Tapscott, 1996).

Second, learner-centered, constructivist approaches are most effective with adult learners when using educational technology (Duffy & Jonassen, 1994). The progress in technology that has brought computers and telecommunications together has finally provided the device that will allow the implementation of John Dewey's progressive educational philosophy. Although Dewey's educational philosophy has been tried repeatedly for almost 100 years, it has been difficult to implement because of the extreme demands on the teacher's time and ability. Now much of this work can be done by computer programs and project-based assignments tailored to the specific needs of the students, set in real-world environments. Constructivism, a learning theory influenced by Dewey's philosophies, can finally be put into practice.

Third, there is an increasing recognition of the importance of work in relationship to education. This change is occurring from two ends. Businesses are recognizing the fact that work is becoming increasingly a mental activity, and therefore a knowledgeable workforce is a key competitive advantage (Fisher & Fisher, 1998). At the same time, higher education is becoming more involved in corporate training and understanding the importance of real-world learning. As the percentage of mental work increases in businesses, and university teaching methods become increasingly reality-based, the role of workplace learning is more central.

The data presented in the two surveys conducted for this book demonstrate overall low levels of sophistication in approaches to distance learning. Universities largely rely on packaged software as platforms for their online courses, regularly defer to technologically inexperienced faculty for course development, and do not exhibit an awareness of the computer medium's special properties. There is much to be done to improve distance learning in American higher education. With the intention of making progress towards improving this state, we now turn to a discussion of those things that are clear about the design of computer-based educational environments. These principles should be understood and used by those responsible for university distance learning programs. I then follow this discussion with a look at some notions of where the integrated use of technology in education might lead us.

WHAT WE KNOW

The following is a list of what I see as clear principles to consider when designing computer software for educational purposes.

Methods

Use Distance Learning Implementation to Reexamine Traditional Teaching/Learning Practices

One of the clear and most intriguing outcomes of the research presented in this book is the way that the often controversial use of distance learning has led to a reexamination of traditional teaching methods. For software designers working in traditional educational administrative environments, this tendency is useful to understand and promote. Bedrock issues for academics, such as contact or seat time, office hours, class scheduling, evaluations of student work, lecture versus discussion, group work, tutorial arrangements, and the repositioning of the faculty member in the classroom, all arise when institutions implement distance learning.

Students in distance learning courses report that their experience was an effective way to learn. The data show that computer-based courses compare favorably with face-to-face courses in terms of learning outcomes, use of critical thinking, and an avoidance of learning primarily by memorization. The implementation of distance learning at a university provides a rare opportunity to freshly examine pedagogical and administrative practices because basic assumptions about traditional learning such as seat time are questioned. Educators should take advantage of this opportunity and think hard—not only about distance learning, but about how some of the practices and principles learned in distance learning can be put to good use in the traditional classroom.

Learning Theory Should Fit Objectives

We saw in Chapter Two that both behavioral and constructivist approaches have advantages for certain kinds of skills to be learned and certain types of students. However, one of the main problems with technology-enabled learning is that it has been used to automate inappropriate behaviorist approaches to certain subjects. Software designers need to understand which approach makes sense given the subject matter, the learners, and the learning objectives. My own personal preference is towards students constructing their own knowledge through the use of computers. I think this is the right approach for working with adults studying complex content. However, I also understand the usefulness of behavioral approaches for more remedial learning objectives and for lower grade levels.

Beware of the Power of Computer-Based Educational Programs to Transmit Ideology

One primary lesson for practitioners that I hope comes out of this book is to understand the potential power of computer-based learning environments as a medium to transmit ideology. We saw how, from the start, both still and moving photography were recognized for their strong ability to persuade viewers of the representation of reality, provide evidence of truthfulness, and to tap into deep psychological structures of the viewer. Regardless of political slant, social realist documentary photographers such as Dorothea Lange and Walker Evans, Nazi filmmakers such as Leni Reifenstahl, Hollywood filmmakers, and American textbook publishers have all utilized the power of photography to transmit ideology. While most designers of the media have taken advantage of the passive positioning of the spectator, some others—such as non-narrative filmmakers, Brecht, and the Surrealists—have specifically tried to prevent this natural tendency. Educators and software designers have a serious responsibility as gatekeepers to the educational process in guarding against the passive transmission of ideology, either through adopting non-illusionist techniques or by providing students with the tools and critical perspective to interpret the media.

Techniques

Communication with Faculty is a Key for Successful Distance Learning Courses

We saw in our research that faculty often reported that teaching distance learning format courses was more difficult and time consuming. Likewise, students reported that often the distance learning courses were loaded with content to compensate for the lack of face-to-face time. Faculty also commented on the difficulty of more subtle forms of communication such as judging student character in online environments. For students, it is clear from the data that the number and quality of the interactions with faculty members is a key to the learning experience's success. Consequently, designers of computer educational environments must pay close attention to emphasizing ways to facilitate faculty-to-student communication.

On the pedagogical level, it is important to examine the number and quality of the communications between the instructor and student in the planning stage. We saw that there are two major functions or roles that faculty members play in distance learning courses. The primary role is that of a tutor, often mostly serving as a resource to answer student questions. One important function of the distance learning faculty member is to help prioritize and direct student attention to specific areas of the often overwhelming amount of content presented. When using the group learning method, faculty also play a key role facilitating and working with the group as it progresses through interdependent tasks. Regular communication with the instructor is vital to achieving success in both of these formats.

Increase Interactivity Among Students

In Chapter Five, I addressed two distinct types of interactivity: student-to-student and teacher, and student-to-media. The 2001 survey showed that students felt the quality of interaction in distance learning courses between the instructor and students was comparable to that found in face-to-face courses. Specifically, 74.6% of the respondents from computer-based courses strongly agreeing and agreed with the statement, "The quality of the interaction with the instructor was the same or better as in traditional face-to-face courses." In terms of student-to-student interaction, the 2001 survey found that there exists slightly higher disagreement (54.3% for computer-based courses; 83.9% for videotape) with the statement "the quality of the interaction with other students in the course was the same or better as in a traditional face-to-face course." Evidently the respondents feel that the quality of the interaction between student and instructor is better than the interaction among students. This finding points to the problem, particularly in courses using a tutorial method, that the important student-to-student communication aspect of interactivity found in the classroom is missing. For some students this loss of student-to-student interaction will make this form of computer-based learning untenable.

According to the 2000 survey, most administrators claim that their courses have a significant amount of interaction among students. Oddly enough, not all distance learning students feel that interactivity is important. According to the results of the 2001 survey, more than half (52.4%) of the respondents did not agree with the statement, "I would prefer active interaction with the course material, instructor, and other students over recorded lectures and prepared materials." While many assume that interactivity is a positive and necessary attribute of effective distance learning courses, some students are drawn to a more solitary and passive learning experience. For this reason, increased student-to-student interaction should probably be voluntary.

For software designers, interactivity is a central concern. While computers offer the opportunity for a great deal more interactivity than passive media such as film and video, how much interactivity should be used? To judge this, designers should consider grade level, learning objectives, and the nature of the subject matter. It is characteristic of the computer medium to provide increased interactivity over other media but, at the same time, also to cause problems with users getting lost in following hypertext links. We saw in the review of media theory that interactivity is absent from most media, yet at the same time, computers incorporate photographic images, moving images that often position viewers in a passive way. Designers need to consider and take advantage of the interactivity provided by computers and beware of the passivity of borrowed media such as film and video.

Use Effective Navigation Schemes

In designing software for students, one of the great challenges is in finding effective ways for users to make their way through often confusing and disorienting environments. Chapter Six looked at navigation issues from both the student and software designer points of view. The 2001 study found 91.9% reported strong agreement or agreement with the statement, "I like maximum control over how to navigate through the course software." In the 2000 study, 84.7% university administrators agreed or strongly agreed with the statement that great care is taken in understanding how students navigate through the computer software. In interviews with the distance learning administrators, this concern for effective computer navigation strategies was also evident. To some extent, the use of narrative and simulations help address the navigation problem, but these approaches are not always appropriate or practical. In these instances, designers need to concentrate on usability and interface design.

Borrow From Older Media

We saw in Chapter Nine that any time a computer educational program employs video or a series of images, it automatically calls up these established film conventions in the student's minds. Designers must be aware of these conventions and in the future develop their own. The hope is that this book will contribute to an understanding of the form those conventions should take. Although differences exist between the media, largely centering on interactivity, the learner already understands the language of film. Designers should take advantage of the learner's ability to understand transitions, implied point of view, and various narrative devices. The second part of the book explored questions of how transition and sequence principles from film might be employed in computer environments. Since viewers are accustomed to the coding of film and television, might not these principles be useful for controlling navigation? While it is inevitable that the computer as a medium will develop its own unique conventions, it can make good use of film and television conventions during this transition phase.

Tutorial or Group Learning Models

We saw in Chapters Three and Four that there are two main approaches to distance learning: group and tutorial. The distinction between these two methods is crucial to understand in both the administration and teaching of distance learning courses, especially in the United States where group teaching methods appears are widespread.

In Chapter Four, group or team learning emerged as the clearest issue of importance in the research literature, which reveals advantages for students working in groups, particularly in online environments. It helps motivate students, builds a sense of community, and increases learning. In support of this notion, we found in the 2001 survey that most distance learning administrators state that their courses offer an opportunity to collaborate with other students. Furthermore, in the interviews from the 2000 study, student collaboration was identified by many administrators as important.

However, the 2001 survey found that many students do not want group interaction. In fact, 73.3% of computer-based course students responded that they either disagreed or strongly disagreed with the statement, "I like to have group projects and other opportunities to learn in group situations." In terms of interdependent tasks, respondents once again to the 2001 survey indicated disinterest in taking part in this type of group learning. In this survey, 63.2% of the computer-based students either disagreed or strongly disagreed with the statement, "I would prefer to communicate with other students around interdependent tasks, rather than open socializing opportunities at a distance." One needs to be conscious of the fact that many of the independent-minded students drawn to this delivery method do not want to be a part of a team.

One position an educator can take is to use the research supporting group learning and insist that students learn in this fashion. This approach is less likely to succeed with adults who generally do not want to be told what is good for them. Another approach would be to only offer tutorial model courses. A third alternative could utilize both methods by making group work an optional but recommended part of the online courses. Regardless, educators need to be aware of these two basic approaches to distance learning and consider which approach to take given both the audience they serve and their educational objectives. On the research end, scholars must understand better how tutorial and group methods function in computer-based educational environments and work towards creating more effective software applications and learning experiences.

Learn Through Programming

One of the more traditional notions of how computers might be used for learning is to teach programming in order to help build conceptualization skills. We saw in Chapter Seven how Seymour Papert sees value in not only having students use computer applications, but in learning how to use programming languages. He emphasizes a constructivist process-oriented approach. By placing themselves inside the symbolic universe of computer programming and trying to move about, students assume a closer relationship to their problem.

While limited, the basic principle of using a computer as a tool to visualize the thinking process offers designers an important model. The goal should be to find a practical way to incorporate this strategy into various subjects and make it an on-going practice, and might be achieved by giving standard online course software the capability for students to do simple programming.

Use Concept Maps

Because it directly involves building cognitive models, the recent research on concept mapping with computers is related to Papert's work. As outlined in Chapter Seven, concepts maps are spatial or graphic representations of concepts and their interrelationships intended to represent human knowledge structures. Concept mapping is important in computer-enabled learning because it puts learners directly in control of the learning experience. As such, it has the potential to become the primary form of desktop interface and consequently change the way humans work and learn with the computer. Furthermore, concept mapping fits with the cognitive research pointing to the human brain as working on a network model. The basic premise behind the use of concept mapping in computers is that by making the structure of thinking graphically explicit, students can better understanding the learning process.

The research literature also shows that concept maps are used to assess student learning. Teachers can gleam from concept maps a good understanding of students' thought processes and thus more easily understand at what point in that process students are having trouble making linkages. Concept mapping can also be used to exchange ideas among students. Designers of educational computer applications should consider how they can either structure interfaces as concept maps, or give students the capability of easily creating and sharing these useful graphical representations of thought processes.

Use Common Computer Applications for Learning

Rather than just concentrating on designing original or new software for educational purposes, designers should consider how widely available software programs might be used for educational purposes. As we saw in Chapter Seven, Jonassen (2000) argues for something he calls "mindtools," or computer-based tools and learning environments that function as intellectual partners with the learner. Unlike other theorists who focus on more exotic futuristic software, Jonassen writes about using current, widely available software such as databases and spreadsheets for educational purposes. He claims that database management tools help learners integrate and interrelate content, and that building databases requires learners to organize information by identifying the content's essential aspects. In designing computer educational environments, designers should look at how they might integrate widely available software in meeting their learning objectives.

Use Simulations

Chapter Twelve showed that the most successful constructivist application of computer-assisted learning has been in the use of simulations. The military, airline pilots, and K-12 educational software, such as the *SimCity* series, have all used simulations successfully. It is inevitable that simulations will be increasingly used in educational environments. The challenge is to create them cost-effectively, and more important, to make sure that they add real value to the educational process. In the early stages of distance learning development, there has been a fascination with the technology, and many universities have spent a great deal of money on creating simulations. Often this money has not been well spent because the simulations really have added little to the educational experience.

Make Learners Aware of Point of View and Subjectivity

In order to prevent uncritical, passive learners, stories should be told from multiple perspectives. The challenge here is to discover how this approach can be achieved most effectively so as not to interfere with linear storytelling's cognitive advantages. Furthermore, designers should incorporate cinematic techniques into interface design that make the learner aware of narrative point of view and the artifice of meaning construction. Offering learners the perspective of different voices in relationship to subject matter in computer environments might be one useful technique.

Limit Voice-Over Narration

As noted in Chapter Ten, sound used in educational software, as in documentary film, is often dominated by narration. For computer-based educational programs, voice-over narration is particularly useful because it helps unify and organize often confusing hypertext-linked content. Nevertheless, we saw in Chapter Nine, documentary film theory warns that voice-over narration works against critical thinking and therefore does not constitute a good learning approach. The voice-over tells the learner how to interpret the images instead of letting him or her draw meaning independently. While an instructor's function through voice-over is to direct attention—an important feature in computer environments—it should not dominate. One way that designers might use voice-over is to delay the voice-over until after the learner has had time to view and interpret in his or her own way.

Make Learners Aware of the Meaning-Making Activity

Chapter Seven established the importance of the strategy of meta-cognition making learners aware of their own thinking and learning processes. Through computer programming, concept maps, and narrative, this meta-cognition activity can be encouraged. In terms of using narrative, a self-conscious approach with notation and feedback mechanisms built into the software would be one practical way of using a meta-cognitive approach.

Create Communities

It is clear that computer software's primary role in online environments is to enable communication among learners and the instructor, as well as to form communities. Working in teams on interdependent tasks is one concrete way the sense of community is encouraged.

Focus on Learning Usability

As indicated in the literature review in HCI (Chapter Five) many design principles should be utilized in educational software's development. First, principles of human factors and usability need to be incorporated in educational software design. Shneiderman's five human factors of time to learn, speed of performance, rate of errors, subjective satisfaction, and retention over time are very useful in education. Furthermore, the overall philosophical approach of usability connects well with learner-centered educational approaches. Finally, the usability emphasis on supporting a range of user skills and needs is essential in today's diverse world.

VISIONS FOR THE FUTURE

The following are some ideas about where the greatest opportunities lie for the use of computers in adult education.

Methods

Increase Informal Learning

The notion of informal learning is extremely important in building a vision for future computer uses. Many fail to recognize that learning occurs outside the classroom—perhaps to a greater extent than in the classroom. Think about how informal learning in daily life might occur, how it might be increased. How can informal learning be encouraged and structured using computers?

Integrate Learning with Work

In the 2001 survey, respondents recognized the importance of incorporating distance learning with work. As shown in Figure 45, 76.7% of the students either strongly agreed or agreed with the statement, "I would like my distance learning to be well integrated with my work/job."

As Chapter Two illustrated, research shows artificial settings and formal training act as a barrier to workplace learning. Learning is most effective when it is informal and takes place in the real workplace environment. Well-integrated learning used to build knowledge and skills in a safe environment using coaching and

Figure 45: Integration of Learning with Work (Questions 1 & 18)

Delivery format * I would like my distance learning to be well integrated with my work/job. Crosstabulation

			I would like my distance learning to be well integrated with my work/job.				Total
			strongly agree	agree	disagree	strongly disagree	
Delivery format	computer-based	Count	21	23	6		50
		% within Delivery format	42.0%	46.0%	12.0%		100.0%
	videotape	Count	6	24	15	2	47
		% within Delivery format	12.8%	51.1%	31.9%	4.3%	100.0%
	correspondence	Count	1	3	1		5
		% within Delivery format	20.0%	60.0%	20.0%		100.0%
	other	Count		1			1
		% within Delivery format		100.0%			100.0%
Total		Count	28	51	22	2	103
		% within Delivery format	27.2%	49.5%	21.4%	1.9%	100.0%

teams may be the most effective. Because computers can be integrated naturally in the workplace in an informal manner, they are well positioned to aid in informal learning.

Present Educational Content in Narrative Form to Increase Comprehension

Chapter Twelve explored how storytelling and the use of narrative serve as powerful tools to transmit cultural meaning, and thereby may be effective in education. As computer-based educational applications evolve, narrative may prove one of the most important methods of making new technology productive for learners.

However, the use of narrative in computer environments is not a simple matter. First, it is a complex undertaking to understand narrative and employ it effectively to increase learning. Second, Chapter Ten cautions that narrative can lead to media user passivity and cause uncritical thinking. The photographic media's strong ability to assert ideological positions, amplified by such images on computers in a narrative format, could be dangerous. Therefore, while it is likely that the use of narrative in computer environments can help navigation and increase learning, it also offers potential dangers if designers proceed uncritically.

Tap Dreams Structures

While it is common for students to drift off into daydreams during classes, no one has ever addressed how this tendency might be harnessed for education. Research that points to the similarities between the internal process of viewing a film and dreaming may relate directly to the experience of learners using computers. If viewers receive films as dreams, narratives might best be structured to take advantage of this predisposition.

Employ Artistic Techniques to Challenge Learners

We looked at the film criticism literature to see how it might inform an understanding of computer-based educational environments. I've described the way films are like dreams, but, in fact, there are many ways in which films are nothing like dreams. Dreams are much more complicated than Hollywood-style films, which often use simple storylines and premises. Moreover, a primary function of dreams, some argue, is to disturb the dreamer—to present material from the unconscious mind that the conscious mind needs to confront. Disturbing an audience is the last thing that a Hollywood producer wants—but could this tactic be useful in education? Non-Hollywood films, most notably as discussed in this book those of the Surrealist Luis Bunuel, are complex, confrontational, and deeply disturbing. While one would not want a learner in a constant state of shock, using structures to provoke and motivate learners could be very useful. This is an artistic strategy rather than a traditional educational one. Nevertheless, particularly for adult learners, this use of artistic provocation might well serve as a sophisticated technique in future educational software.

Accommodate Learning Styles

As reviewed in Chapter Two, the extensive body of research on learning styles may provide key insights for customizing educational software. In a regular classroom, teachers are forced to use generic approaches to learning style; however, with the use of special software in distance learning, teachers might be able to customize the learning experience using those methods best suited to the style of the student. The challenge here is to find ways that the computer can identify and then adapt to the user's learning style. At this time, neither the research on learning styles nor the technology to accommodate those styles has advanced far enough for this approach to work. However, in the future both of these agendas will have advanced far enough to begin to accommodate learning styles and thereby increase learning, at least on a simple level.

Techniques

Visualization

We saw in Chapter Seven that the research literature often describes the use of the computer for visualization and creative endeavors. Confirming the desire for this kind of functionality, in the 2001 survey we found that students had a clear interest in using the computer for visualization and creative endeavors: 89.1% of the respondents indicated they either strongly agree or agree with the statement, "I would like to use computers for creative and visualization purposes in distance learning programs."

Computer designers building programs for educational purposes need to consider how they can provide these visualization capabilities to enhance creative thinking. How might this be done? One of the distinct advantages of computers is the ability to create and combine images, as well as text, quickly. If it is structured into a computer tool, users can take advantage of this ability to help spark creative thinking. More broadly, this visualization capability already occurs, or at least to some degree, through word processors, spreadsheets, and graphic programs. Word processors alone have revolutionized how people write by allowing on-the-fly editing and creating of documents. In expanding this analogy to other fields, one can see that the computer might very well provide excellent tools to assist and spark creative thinking.

Intelligent Tutors or Agents

Chapter Three explored one of the most exciting tools for learning: the labor-saving intelligent tutors or agents. As research assistants, intelligent agents might track the user's tendencies, then collect information that fits his or her specific interests. While it may be true that artificial intelligence will never be able to replicate a human teacher completely, intelligent agents customized to fit the user's interests may well serve as useful tools approximating some tutorial functions.

Give Learners the Opportunity to Re-create Their Own Story That Incorporates External Narrative Material with Their Own Personal Knowledge

In Chapter Twelve we learned about clinical uses of narrative psychology that involve clients constructing their own narratives. Can Michael White's notion of re-authoring in clinical practice work in the design of educational programs? Computers might be used to allow students to re-author the stories from which they have learned and thereby actively build their own meaning structures based on the course content. An example of this method might be to give learners a list of key or random words and images from a text, then ask them to either build a narrative or write a letter. With their ability to easily manipulate text, sound, and images, computers could be powerful tools in this personal story-building process.

LEARNING DEVICES

Now that I have described the elements of how we might learn, I want to look towards the near future and identify the methods and technologies that should be utilized. If we were to design a lifelong learning device best suited to create and maintain virtual learning communities, what would it look like? What particular characteristics would be central to its design?

Perhaps this device's most important function would be to enable the successful formation of learning communities. On an individual level, the device would constantly manage and update the various communities to which an individual might belong. It might alert a user to new learning communities and suggest membership in certain communities. Learning devices might also recognize the need for specific new communities and help form them by contacting potential members based on user profiles. In virtual learning communities, communication is a key function. In the future, those institutions that understand their role in providing this setting for communication will be the most successful.

Furthermore, it is very important for learning communities to recognize patterns of meaning and enable both individual and group process memory. In 1945, Vannevar Bush sketched out a plan for something very much like what is known today as hypertext. Bush's article points out that the scientific world is overrun with too much information; in order to progress, man needs to develop technology to address this problem. Calling his device "memex," Bush argued that it should use association because that method mirrors the way humans think. The problem to overcome is that these associations are very intricate and leave quickly fading trails. Part of what this device does is link documents together, much like hypertext ultimately generated. However, one important aspect of Bush's device has not been replicated—the ability to have a memory of thinking, or "thinking trails" (Johnson, 1997). Bush felt that this ability to capture thinking trails was crucial in future thinking devices so that individuals and groups could trace connections and the branching of thinking in order to examine their own processes and learn from them. In addition to the significance of this process orientation in learning communities, the memory of this process also emerges as key. As pointed out in Chapter Four, studies of virtual teams have shown that this process memory is one of the common problems encountered (Weiser & Morrison, 1998). Often team members working on different tasks lose the thread of inquiry, and such a device to help track group thought process is posited by Weiser and Morrison. This kind of process memory is just what Bush was writing about and is the kind of thing that is needed in virtual learning communities. Bush's essay points out that the recognition of patterns or thinking trails is a key need in future thinking devices. The sharing of thinking patterns among virtual learning communities would be a central practice. Computers can accomplish pattern recognition and memory very efficiently. Lifelong learning devices would chart thinking trails, suggest possible connections with other trails, and help share patterns with other learners.

Another function of the lifelong learning device is the ability to help structure work-based learning. This task would be accomplished by a device that connects knowledge trails produced outside of work with the information needed to accomplish specific employment-related tasks. The device would suggest relevant information sources outside of work, virtual learning communities, and thinking trails that could have an immediate impact on a user's work.

The lifelong learning device would most likely have a software component and an input device. Bush makes reference to an input device in his future thinking machine to record various types of information, including photographs and print material. While a great deal of the information needed for this device would already be digitized, in order to make the information personal and particular, the user would require the ability to input information. The information held in the device would need to be accessible anywhere. Much like the Internet, the user would need to be able to access, personalized learning device at work, school, home and in transit. The lifelong learning device would be something that an individual keeps for a lifetime, and it would go with him or her everywhere. As a result, a learner's collection of teams would also be perpetually accessible.

The device must be dynamic and continually in the process of updating information for the user. This function is likely to include a certain amount of artificial intelligence able to understand the user's needs and interests and find the necessary information. Much of this could be accomplished by the intelligent agents that are currently being discussed and developed.

The interface of this lifelong learning device should be based on purpose rather than current metaphors. Why organize around space instead of meaning? Metaphors and virtual university icons fail to center the interface on the real learning process. Traditional university metaphors are particularly unproductive because they do not focus on individual student needs and thought processes. Most important, the interface should reflect a feeling of a community of people—not an individual, private desktop.

Continual customization will be part of the lifelong learning device in order to meet individual learner and team needs on an ongoing basis. Simulation has been one of the most successful training tools thus far developed. Military uses, flight simulation, and successful popular games such as *SimCity* and *Civilization* are examples of early successes. Future learning devices are likely to include simulation and role-playing opportunities, particularly in team environments in which learners can experience the simulations together.

Random connections suggested by computers and other learners emerge as another important characteristic of learning machines. Using simple programmed randomness, computers can suggest connections between thoughts that will serve as a powerful learner brainstorming tool.

RESEARCH AGENDA

Whether constructivist, behaviorist, or working from another theoretical paradigm, practical theoretical applications through specific techniques that get results are going to ultimately be the most convincing and useful. While this book's scope is too broad to explore any one technique in detail, I can sketch out some basic directions where future research is likely to go. The following are some areas for further research:

1) Verify the power of narrative in learning;
2) Test the results of retelling from multiple perspectives;
3) Develop ways to adapt to learning styles;
4) Investigate the use of dreams as a narrative structure;
5) Look at ways to emphasize community creation;
6) Create tools and devices for cognitive amplification;
7) Test how moving and still photographs are perceived by learners;
8) Test artistic methods to challenge and provoke thinking; and
9) Investigate how to increase informal learning with computers.

Many of these primary research questions involve the testing of principles learned from related media such as narrative point of view and, especially, editing—a key feature of meaning making in computer educational environments. As we saw in the second part of this book, many of the principles of film editing are relevant to the development of computer educational environments, including: context (Kuleshov), parallel montage (Eisenstein, Griffith), and more specifically, issues of sequence length, use of transitions, and camera movement.

FINAL THOUGHTS

Though in many ways revered, the physical classroom may be what has weighed down visionary notions of education. Senge (2000) comments on this notion of a need for greater integration of the human growth occurring in schools with the rest of life:

> The classroom itself may be the fundamental limit in recreating higher education. The classroom reinforces a teacher-centered view of education. It easily becomes the stage for confusing teaching and learning. Moreover, it is a symbol of the overall isolation for the university from the larger world (Senge, 2000).

One can argue that there may be wisdom separating the university from the larger practical world so that society receives benefits of research indirectly. However, it is difficult to validate that human change brought about by education on an individual level should occur only in formal university settings. Finally, computer-based learning is not about "convenience" as many claim, but rather about providing more informal and integrated opportunities to learn. Thus, the lesson for educational software designers is to create learning environments that emphasize this integrated and natural way of learning.

To my mind, the challenge for educators and software designers is to move beyond the automation of the traditional classroom. Ideally, they should aim to create a new future where learning is richer, deeper, and more integrated with daily life through the use of this marvelous new knowledge medium: the personal computer.

References

Abelson, P. H. (1997). Evolution of higher education. *Science Magazine, 277*(5327), Issue 8.

Abrahamson, C. E. (1998). Storytelling as a pedagogical tool in higher education. *Education, 18*(3), 440.

Adler, P. S. and Winograd, T. A. (1992). *Usability: Turning Technologies Into Tools.* Oxford: Oxford University Press.

Altman, R. (1992). *Sound Theory Sound Practice.* New York: Routledge.

Aretz, A. J., Bolen, T. and Devereux, K. E. (1997). Critical thinking assessment of college students. *Journal of College Reading and Learning.* 27(Fall).

Argyris, C. and Schon, D. A. (1974). *Theory in Practice: Increasing Professional Effectiveness.* San Francisco, CA: Jossey-Bass Publishers.

Aristotle. (1941). *The Basic Works of Aristotle.* New York: Random House.

Arnheim, R. (1974). *Art and Visual Perception: The Psychology of the Creative Eye.* Berkeley, CA: University of California Press.

Baecker, R. and Small, I. (1995). Animation at the interface. In Laurel, B. (Ed.). *The Art of Human-Computer Interface Design.* Reading, MA: Addison-Wesley Publishing Company.

Baker, M. Hanse, T, Joiner, R. and Traum, D. (1999). The role of grounding in collaborative learning tasks. In Dillenbourg, P. (Ed.), *Collaborative Learning: Cognitive and Computational Approaches.* New York: Pergamon.

Barsam, R. M. (1992). *Non-Fiction Film: A Critical History.* Bloomington, IN: Indiana University Press.

Baskett, H. K. (1993). Workplace factors which enhance self-directed learning. Presented to the *Seventh International Symposium on Self-Directed Learning,* January 21-23, 1993, West Palm Beach, FL.

Baskett, H. K. and Morris, D. (1992). Conditions enhancing self-direct learning in the workplace. *Social Sciences and Humanities Research Council of Canada.* July. (ERIC ED 352 563).

Bederson, B. B. (1998). *Does Animation Help Users Build Mental Maps of Spatial Information?* Computer Science Department, Human-Computer Interaction Lab, University of Maryland. Unpublished.

Berg, G. A. (2002). *Why Distance Learning? Administrative Practices in Higher Education*. Phoenix, AZ: Onyx Press.

Berger, J. (1972). *Ways of Seeing*. New York: Pelican Books.

Bernard, H. R. (1988). *Research Methods in Cultural Anthropology*. Newbury Park, CA: Sage Publications.

Black, J. B., Kay, D. S. and Soloway, E. M. (1987). Goal and plan knowledge representation. In Carroll, J. M. (Ed.), *Interfacing Thought: Cognitive Aspects of Human-Computer Interaction*. Cambridge, MA: The MIT Press.

Bordwell, D. (1989). *Making Meaning*. Cambridge, MA: Harvard University Press.

Bordwell, D. (1997). *On the History of Film Style*. Cambridge, MA: Harvard University Press.

Bourdeau, J. (1999). The problem of transitions between discrete multimedia spaces. *Proceedings of EDM 99*. Charlottesville, VA: AACE.

Branigan, E. (1984). *Point of View in the Cinema*. Berlin, New York, Amsterdam: Mouton Publishers.

Brookfield, S. D. (1986). *Understanding and Facilitating Adult Learning*. San Francisco, CA: Jossey-Bass.

Brookfield, S. D. (1987). *Developing Critical Thinkers*. San Francisco, CA: Jossey-Bass Publishers.

Brooks, K. M. (1999). *Metalinear Cinematic Narrative: Theory, Process, and Tool*. Unpublished paper.

Brownlow, K. (1968). *The Parade's Gone By*. Berkeley, CA: University of California Press.

Brownlow, K. and Kobal, J. (1979). *Hollywood: The Pioneers*. New York: Alfred A. Knopf.

Bruffee, K. (1995). Sharing our toys—Cooperative learning versus collaborative learning. *Change*, (January-February), 12-18.

Bruner, J. (1986). *Actual minds, Possible Worlds*. Cambridge, MA: Harvard University Press.

Bruner, J. (1990). *Acts of Meaning*. Cambridge, MA: Harvard University Press.

Bruner, J. (1996). *The Culture of Education*. Cambridge, MA: Harvard University Press.

Burch, N. (1981). *Theory of Film Practice*. Princeton, NJ: Princeton University Press.

Bush, V. (1945). As we may think [electronic version]. *The Atlantic Monthly*, July.

California Post-secondary Education Commission (CPEC). (1997). *Coming of Information Age in California Higher Education*. Sacramento, CA.

Campbell, D. T. and Stanley, J. C. (1966). *Experimental and Quasi-Experimental Designs for Research*. Chicago, IL: Rand McNally.

Campbell, S. F. (1976). *Piaget Sampler*. New York: John Wiley & Sons.

Card, S. K., Moran, T. P. and Newell, A. (1983). *The Psychology of Human-Computer Interaction*. Hillsdale, NJ: Lawrence Erlbaum Associates, Publishers.

Carey, J. M. (Ed). (1991). *Human Factors in Information Systems: An Organizational Perspective*. Norwood, NJ: Ablex Publishing Company.

Carroll, N. (1988). *Mystifying Movies: Fads and Fallacies in Contemporary Film Theory*. New York: Columbia University Press.

Chan, T. S. and Ahern, T. C. (1999). The importance of motivation: Integrating flow theory into instructional design. *Proceedings of SITE 99*. Charlottesville, VA: AACE.

Chapman, D. (1991). *Vision, Instruction, and Action*. Cambridge, MA: The MIT Press.

Chatman, S. (1978). *Story and Discourse: Narrative Structure in Fiction and Film*. Ithaca and London: Cornell University Press.

Chiu, C., Chen, H., Wei, L. and Hu, H. (1999). Approaching effective network cooperative learning. *Proceedings of PR 99*. Charlottesville, VA: AACE.

Cohen, A. D. (1997). *The Value of Metaphors in Conceptual User-Interface Design. Unpublished dissertation*. Northwestern University.

Confessore, G. J. (1996). *Consideration of Selected Influences on Work Place Learning*, May. (ERIC ED 401 420).

Cook, D. L. (1995). Community and computer-generated distance learning environments. In Rossman, M. H. and Rossman, M. E. (Eds.), *Facilitating Distance Education*. San Francisco, CA: Jossey-Bass Publishers.

Cook, T. D., Appleton, H., Conner, R. F., Shaffer, A., Tamkin, G. and Weber, S. (1975). *"Sesame Street" Revisited*. New York: Russell Sage Foundation.

Covey, S. R. (1995). The principle of continuous learning. *Executive Excellence*, April.

Cross, K. P. (1981). *Adults as Learners*. San Francisco, CA: Jossey-Bass Publishers.

Csikszentmihalyi, M. (1991). *Flow: The Psychology of Optimal Experience*. New York: HarperCollins.

Cuban, L. (1986). *Teachers and Machines: The Classroom Use of Technology Since 1920*. New York: Teachers College Press.

Cusimano, J. M. (1995). Turning blue-collar workers into knowledge workers. *Training & Development*, August.

Daniel, J. (1997). *Mega-Universities and Knowledge Media*.

Darragh, J. J. & Witten, I. H. (1992). The reactive keyboard. Cambridge: Cambridge University Press.

Davenport, G., Agamanolis, S., Bradley, B. and Sparacino, F. (1997). Encounters in dreamworld: A work in progress. Presented at *Consciousness Reframed, 1st International CAiiA Research Conference*, University of Wales College, Newport, Wales, July.

Davis, S. and Botkin, J. (1994). *The Monster Under the Bed.* New York: Touchstone.

DeMartino, D. J. (1999). Employing adult education principles in instructional design. *Proceedings of SITE 99.* Charlottesville, VA: AACE.

DeVaney, A. (1990). Instructional television without educators: The beginning of ITV. In Ellsworth, E. and Whatley, M. H. (Eds.), *The Ideology of Images in Educational Media.* New York: Teacher's College Press.

Derrida, J. (1973). *Speech and Phenomena.* Evanstron, IL: Northwestern University Press.

Dertouzos, M. (1997). *What Will Be.* New York: HarperCollins.

Deslauriers, D. (1992). Dimensions of knowing: Narrative, paradigm, and ritual. *ReVision*, Spring.

Dewey, J. (1927). *The Public and Its Problems.* Denver, CO: A. Swallow.

Dillenbourg, P. (Ed). (1999). *Collaborative Learning: Cognitive and Computational Approaches.* New York: Pergamon.

DiPaolo, T. (1999). Learning strategies: A framework for understanding students learning with computers. *Proceedings of EDM 99.* Charlottesville, VA: AACE.

DiSessa, A. A. (2000). *Changing Minds: Computers, Learning, and Literacy.* Cambridge, MA: MIT Press.

Duffy, T. M., Dueber, B. and Hawley, C. L. (1998). Critical thinking in a distributed environment: A pedagogical base for the design of conferencing systems. In Bonk, C. J. and King, K. S. (Eds.), *Electronic Collaborators: Learner-Centered Technologies for Literacy, Apprenticeship, and Discourse.* Mahwah, NJ: Lawrence Erlbaum Associates.

Duffy, T. M. and Jonassen, D. H. (1994). *Constructivism and the Technology of Instruction: A Conversation.* Charlottesville, VA: American Society for Advancement in Computer Education.

Egri, L. (1946). *The Art of Dramatic Writing.* New York: Simon & Schuster.

Eisenstadt, M. (1995). The knowledge media generation. *The Times Higher Education Supplement, Multimedia Section*, pp. vi-vii, April 7. http://kmi.open.ac.uk/kmi-feature.html.

Eisenstadt, M. and Vincent, T. (Eds.). (2000). *The Knowledge Web: Learning and Collaborating on the Net.* London: Kogan Page.

Eisenstein, S. M. (1969). *Film Form*. Edited and translated by Jay Leyda. New York: Harcourt Brace.

Eisentstein, E. L. (1979). *The Printing Press as an Agent of Change: Communications and Cultural Transformations in Early Modern Europe*. Cambridge, MA: Cambridge University Press.

Ellsworth, E. (1990). Educational films against critical pedagogy. In Ellsworth, E. and Whatley, M. H. (Eds.), *The Ideology of Images in Educational Media*. New York: Teacher's College Press.

Ellsworth, E. and Whatley, M. H. (Eds.). (1990). *The Ideology of Images in Educational Media*. New York: Teacher's College Press.

Erdman, B. (1990). The closely guided viewer: Form, style, and teaching in the educational film. In Ellsworth, E. and Whatley, M. H. (Eds.), *The Ideology of Images in Educational Media*. New York: Teacher's College Press.

Erickson, M., Rossi, E. and Rossi, S. (1976). *Hypnotic Realities*. New York: Irvington Publishers.

Etzioni, A. and Etzioni, O. (1997). Communities: Virtual vs. real. *Science*, July 18, 295.

Faulkner, C. (1998). *The Essence of Human-Computer Interaction*. New York: Prentice Hall.

Feifer, R. G. (1994). Cognitive issues in the development of multimedia learning systems. In Reisman, S. (Ed.), *Multimedia Computing: Preparing for the 21st Century*. Hershey, PA: Idea Group Publishing.

Fisher, K. and Fisher, M. D. (1998). *The Distributed Mind*. New York: American Management Association.

Fox, B. A. (1993). *The Human Tutorial Dialogue Project: Issues in the Design of Instructional Systems*. Hillsdale, NJ: Lawrence Erlbaum Associates.

Frasson, C. and Gauthier, G. (1990). *Intelligent Tutoring Systems: At the Crossroads of Artificial Intelligence and Education*. Norwood, NJ: Ablex Publishing Corporation.

Freud, S. (1963). *Character and Culture*. New York: Collier.

Gardner, H. (1973). *The Quest for Mind*. New York: Alfred A. Knopf.

Gardner, H. (1993). *Multiple Intelligences*. New York: Basic Books.

Gibbons, A. S. and Fairweather, P. G. (1998). *Computer-Based Instruction: Design and Development*. Englewood Cliffs, NJ: Educational Technology Publications.

Gidal, P. (1989). Theory and definition of structural/materialist film. In Gidal, P. (Ed.), *Structural Film Anthology*. London: Routledge.

Gilbert, J. E. and Han, C. Y. (1999). Arthur: An adaptive instruction system based on learning styles. *Proceedings of PR 99*. Charlottesville, VA: AACE.

Ginsburg, H. P. and Opper, S. (1988). *Piaget's Theory of Intellectual Development*. Englewood Cliffs, NJ: Prentice Hall.

Goldman, A. I. (1999). *Knowledge in a Social World*. Oxford: Clarendon Press.

Goldman, F. and Burnett, L. R. (1971). *Need Johnny Read? Practical Methods to Enrich Humanities Courses Using Films and Film Study*. Dayton, OH: Pflaum.

Gorden, R. L. (1975). *Interviewing: Strategy, Techniques, and Tactics*. Homewood, IL: The Dorsey Press.

Haber, N. and Hershenson, M. (1973). *The Psychology of Visual Perception*. New York: Holt, Rinehart and Winston, Inc.

Hansen, T., Dirckinck-Holmfeld, L., Lewis, R. and Rugelj, J. (1999). Using telematics for collaborative knowledge construction. In Dillenbourg, P. (Ed.), *Collaborative Learning: Cognitive and Computational Approaches*. New York: Pergamon.

Harre, R. and Gillett, G. (1994). *The Discursive Mind*. Thousand Oaks, CA: Sage Publications.

Hawkins, J. M. (1986). *The Oxford Reference Dictionary*. Oxford: Clarendon Press.

Head, A. (1999). *Design Wise: A Guide for Evaluating the Interface Design of Information Resources*. Medford, NJ: Cyberage Books.

Healey, J. F. (1999). *Statistics: A Tool for Social Research*. Belmont, CA: Wadsworth Publishing Company.

Hedges, I. (1991). *Breaking the Frame: Film Language and the Experience of Limits*. Bloomington, IL: Indiana University Press.

Heflich, D. A. and Rice, M. L. (1999). Online survey research: A venue for reflective conversation and professional development. *Proceedings of SITE 99*. Charlottlesville, VA: Association for the Advancement of Computers in Education.

Hegel, G. W. R. (1956). *The Philosophy of History*. New York: Dover Publications.

Heidegger, M. (1962). *Being and Time*. New York: Harper & Row.

Helander, M. (Ed.). (1998). *Handbook of Human-Computer Interaction*. Amsterdam: North-Holland.

Henderson, B. (1976). Two types of film theory. In Nichols, B. (Ed.), *Movies and Methods*. Berkeley, CA: UC Press.

Heterick Jr., R. C. (1993). Reengineering teaching and learning in higher education: Sheltered groves, camelot, windmills, and malls. *Cause*. Professional Paper Series, #10.

Hoad, T. F. (1986). *The Concise Oxford Dictionary of English Etymology*. Oxford: Oxford University Press.

Hoban, C. F. (1942). *Focus on Learning: Motion Pictures in the School*. Washington DC: American Council on Education.

Huhns, M. N. and Singh, M. P. (Eds.). (1998). *Readings in Agents*. San Francisco, CA: Morgan Kaufmann Publishers.

Innis, H. A. (1972). *Empire and Communications*. Toronto: University of Toronto Press.

Innis, H. A. (1991). *The Bias of Communication*. Toronto: University of Toronto Press.

Jarvenpaa, S. L., Knoll, K. and Leidner, D. E. (1998). Is anybody out there? Antecedents of trust in global virtual teams. *Journal of Management of Information Systems*, Spring.

Johnson, S. (1997). *Interface Culture: How New Technology Transforms the Way We Create and Communicate*. San Francisco, CA: HarperEdge.

Joiner, R. Issroff, K. and Demiris, J. (1999). Comparing human-human and robot-robot interactions. In Dillenbourg, P. (Ed.), *Collaborative Learning: Cognitive and Computational Approaches*. New York: Pergamon.

Jonassen, D. H. (2000). *Computers as Mindtools for Schools*. Englewood Cliffs, NJ: Merrill.

Jonassen, D. H., Tessmer, M. and Hannum, W. H. (1999). *Task Analysis Methods for Instructional Design*. Mahwah, NJ: Lawrence Erlbaum Associates.

Jonassen, D. H., Reeves, T. C., Hong, N., Harvey, D. and Peters, K. (1997). Concept mapping as cognitive learning and assessment tools. *Journal of Interactive Learning Research*, 8.

Joyce, M. (2000). Othermind-ness: The emergence of network culture. Ann Arbor, MI: University of Michigan Press.

Karlin, F. (1994). *Listening to Movies: The Film Lover's Guide to Film Music*. New York: Schirmer Books.

Katz, S. D. (1991). *Film Directing: Shot By Shot, Visualizing From Concept to Screen*. Los Angeles, CA: Michael Wiese Productions.

Kaufman, J. and Goldstein, L. (1976). *Into Film*. New York: Dutton & Co.

Kay, A. (1995). User interface: A personal view. In Laurel, B. (Ed.), *The Art of Human-Computer Interface Design*. Reading, MA: Addison-Wesley Publishing Company.

Kent, T. W. and McNergney, R. F. (1999). *Will Technology Really Change Education?* Thousand Oaks, CA: Corwin Press.

Kerr, C. (1963). *The Uses of the University*. Cambridge, MA: Harvard University Press.

King, K. S. (1998). Designing 21st-century educational networlds: structuring electronic social spaces. In Bonk, C. J. and King, K. S. (Eds.), *Electronic Collaborators: Learner-Centered Technologies for Literacy, Apprenticeship, and Discourse*. Mahwah, NJ: Lawrence Erlbaum Associates.

Kirkley, S. E., Savery, J. R. and Grabner-Hagen, M. M. (1998). Electronic teaching: Extending classroom dialogue and assistance through e-mail communication. In Bonk, C. J. and King, K. S. (Eds.), *Electronic Collaborators: Learner-Centered Technologies for Literacy, Apprenticeship, and Discourse*. Mahwah, NJ: Lawrence Erlbaum Associates.

Kitao, K. (1995). *The History of Language Laboratories: Origin and Establishment* (ERIC ED381020).

Klapper, J. T. (1960). *The Effects of Mass Communication*. Boston, MA: Free Press.

Klima, G. (1974). *Multi-Media and Human Perception*. New York: Meridian Press.

Knowles, M. (1979). *The Adult Learner: A Neglected Species*. Houston, TX: Gulf Publishing Company.

Lacoue-Labarthe, P. (1979). *Theatrum Analyticium*. Glph. #2.

Laurel, B. (1990). *The Art of Human-Computer Interface Design*. Reading, MA: Addison-Wesley.

Laurel, B. (1993). *Computers as Theatre*. Reading, MA: Addison-Wesley.

Laurillard, D. (1997). *Multimedia and the Learner's Experience of Narrative*. British Open University. London: Meno Publication.

Lennon, J. A. (1997). *Hypermedia Systems and Applications: World Wide Web and Beyond*. Berlin: Springer.

Lipnack, J. and Stamps, J. (1997). *Virtual Teams: People Working Across Boundaries With Technology*. New York: John Wiley and Sons.

Littleton, K. and Hakkinen, P. (1999). Learning together: Understanding the processes of computer-based collaborative learning. In Dillenbourg, P. (Ed.), *Collaborative Learning: Cognitive and Computational Approaches*. New York: Pergamon.

Lyotard, J. F. (1992). *The Libinal Economy*. Bloomington, IN: Indiana University Press.

Lynch, W. (1998). Communications technology and video production: An evolutionary study of their effects on a distance learning program. *Proceedings of EDM 98*, 910-915. Charlottesville, VA: AACE.

Maddix, F. (1990). *Human-Computer Interaction: Theory and Practice*. New York: Simon & Schuster.

Malloch, M., Cairns, L. and Hase, S. (1998). Learning in the workplace: Implications of the capability learning model. *Paper presented at the Annual Meeting of the American Educational Research Association*, April. (ERIC ED 418 330).

Malone, T. W. (1987). Computer support for organizations: Toward an organizational science. In Carroll, J. M. (Ed.), *Interfacing Thought: Cognitive Aspects of Human-Computer Interaction*. Cambridge, MA: The MIT Press.

Maslow, A. H. (1970). *Motivation and Personality*. New York: Harper & Row.

McCarter, J. (1996). Learning communities absent from higher education. *The Chronicle of Higher Education*, October 3.

McKnight, C., Dillon, A. and Richardson, J. (1991). *Hypertext in Context*. Cambridge, MA: Cambridge University Press.

McLeod, J. (1996). The emerging narrative approach to counseling and psychotherapy. *British Journal of Guidance & Counseling*, June.

McLuhan, M. (1964). *Understanding Media: The Extension of Man*. New York: McGraw-Hill.

Merriam, S. B. and Clark, M. C. (1991). *Lifelines: Patterns of Work, Love, and Learning in Adulthood*. San Francisco, CA: Jossey-Bass.

Metz, C. (1982). *The Imaginary Signifier: Psychoanalysis and the Cinema*. Bloomington, IN: Indiana University Press.

Meyrowitz, J. (1986). *No Sense of Place: The Impact of Electronic Media on Social Behavior*. Oxford: Oxford University Press.

Mitry, J. (1997). *The Aesthetics and Psychology of the Cinema*. Bloomington, IN: Indiana University Press.

Montgomery, J. R. (1996). Integrative learning at work: Theory into practice at Andersen Consulting. Paper presented at the *Conference of the Academy of Human Resource Development*, February (ERIC ED 399 417).

Montgomery, J. R. and Lau, C. C. (1996). Integrating work and learning for superior performance. Paper presented at the *Annual Conference of the Academy of Human Resource Development*, March 2. (ERIC ED 399 418).

Morris, J. M., Owen, G. S. and Fraser, M. D. (1994). Practical issues in multimedia user interface design for computer-based instruction. In Reisman, S. (Ed.), *Multimedia Computing: Preparing for the 21st Century*. Hershey, PA: Idea Group Publishing.

Mountford, S. J. (1995). Tools and techniques for creative design. In Laurel, B. (Ed.), *The Art of Human-Computer Interface Design*. Reading, MA: Addison-Wesley Publishing Company.

Mulvey, L. (1996). *Fetishism and Curiosity*. London: British Film Institute.

Mumford, L. (1934). *Technics and Civilization*. New York: Harcourt, Brace and Company.

Murray, J. H. (1997). *Hamlet on the Holodeck: The Future of Narrative in Cyberspace*. New York: Free Press.

Nelson, T. H. (1995). The right way to think about software design. In Laurel, B. (Ed.), *The Art of Human-Computer Interface Design*. Reading, MA: Addison-Wesley Publishing Company.

Newby, T. J., Stepich, D. A., Lehman, J. D. and Russell, J. D. (1996). *Instructional Technology for Teaching and Learning*. Englewood Cliffs, NJ: Merrill.

Nichols, B. (1994). *Blurred Boundaries: Questions of Meaning in Contemporary Culture*. Bloomington, IN: Indiana University Press.

Nielsen, J. (1990). *Hypertext & Hypermedia*. Boston, MA: Academic Press, Inc.

Nizhny, V. (1962). *Lessons with Eisenstein*. New York: Hill and Wang.

Norman, D. A. (1987). Cognitive engineering—cognitive science. In Carroll, J. M. (Ed.), *Interfacing Thought: Cognitive Aspects of Human-Computer Interaction*. Cambridge, MA: The MIT Press.

Norman, D. A. (1988). *The Design of Everyday Things*. New York: Currency Doubleday.

Norman, D. A. (1998). *The Invisible Computer*. Cambridge, MA: MIT Press.

Novak, J. D. (1998). *Learning, Creating, and Using Knowledge*. Hillsdale, NJ: Lawrence Earlbaum.

Oren, T. (1995). Designing a new medium. In Laurel, B. (Ed.), *The Art of Human-Computer Interface Design*. Reading, MA: Addison-Wesley Publishing Company.

Oren, T., Salomon, G., Kreitman, K. and Abbe, D. (1995). Guides: Characterizing the interface. In Laurel, B. (Ed.), *The Art of Human-Computer Interface Design*. Reading, MA: Addison-Wesley Publishing Company.

Orner, M. (1990). Open-ended films, dead-end discussions: An ideological analysis of trigger films. In Ellsworth, E. and Whatley, M. H. (Eds.), *The Ideology of Images in Educational Media*. New York: Teacher's College Press.

Osgood, R. E. (1994). *The Conceptual Indexing of Conversational Hypertext. Unpublished dissertation*.

Palloff, R. M. and Pratt, K. (1999). *Building Learning Communities in Cyberspace*. San Francisco, CA: Jossey-Bass Publishers.

Panitz, T. and Panitz, P. (1998). Encouraging the use of collaborative learning in higher education. In Forest, J. J. (Ed.), *Issues Facing International Education*. New York: Garland.

Peters, J. M., Jarvis, P. and Associates. (1991). *Adult Education*. San Francisco, CA: Jossey-Bass Publishers.

Petre, M., Carswell, L., Price, B. and Thomas, P. (2000). Innovations in large-scale supported distance teaching: Transformation for the Internet, not just translation. In Eisenstadt, M. and Vincent, T. (Eds.), *The Knowledge Web: Learning and Collaborating on the Net*. London: Kogan Page.

Piaget, J. (1995). *Sociological Studies*. London: Routledge.

Pickover, C. A. (1991). *Computers and the Imagination*. New York: St. Martin's Press.

Pintrich, P. R. (1995). *Understanding Self-Regulated Learning*. San Francisco, CA: Jossey-Bass Publishers.

Pittman, V. (1986). Pioneering instructional radio in the U.S.: Five years of frustration at the University of Iowa, 1925-1930. *A Paper for the First International Conference on the History of Adult Education*, Oxford, UK, July 14-17, ERIC ED297104.

Plantinga, C. (1994). Pleasures and the spectator's experience: Towards a cognitive approach. *Film and Philosophy, II*.

Plowman, L. (1994). The 'primitive mode of representation' and the evolution of interactive multimedia. *Journal of Educational Multimedia and Hypermedia*, 3(3/4), 275-293.

Plowman, L. (1996). *Getting Side-Tracked: Cognitive Overload, Narrative, and Interactive Learning Environments*. British Open University. London: Meno Publication.

Plowman, L. (1997). Narrative, interactivity and the secret world of multimedia. British Open University. London: Meno Publication.

Preece, J. and Shneiderman, B. (1995). *Survival of the Fittest: The Evolution of Multimedia User Interfaces. Unpublished*.

Rabinowitz, P. (1994). *They Must Be Represented: The Politics of the Documentary*. New York: Verso.

Reason, D. and Forrester, M. (1997). Kinds of narrative, designs of hypermedia. Paper presented at *Meno workshop on Narrative and Hypermedia*, April. London: British Open University.

Reason, J. (1990). *Human Error*. Cambridge, MA: Cambridge University Press.

Reeves, T. C., Laffey, J. M. and Marlino, M. R. (1996). New approaches to cognitive assessment in engineering education. Paper presented at the *Annual Meeting of the American Educational Research Association*, April 8-12. Charlottesville, VA: AACE.

Reilly, A. (1999). Reading and listening: Issues in the use of displayed text and recorded speech in educational multimedia. *Proceedings of EDM 99*. Charlottesville, VA: AACE.

Reisner, P. (1987). Discussion: HCI, what is it and what research is needed? In Carroll, J. M. (Ed.), *Interfacing Thought: Cognitive Aspects of Human-Computer Interaction*. Cambridge, MA: The MIT Press.

Rodowick, D. N. (1980). Vision, desire, and the film-text. *Camera Obscura*, #6.

Rogers, C. R. (1968). *Freedom to Learn*. Columbus, OH: Charles E. Merrill Publishing Company.

Rourke, L., Anderson, T., Garrison, D. R. and Archer, W. (1999). Assessing social presence in asynchronous, text-based computer conferencing. *Journal of Distance Education*, 14(3), 51-70.

Russell, T. L. (1999). *The No Significant Difference Phenomenon*. Chapel Hill, NC: North Carolina State University Press.

Saettler, P. (1990). *The Evolution of American Educational Technology.* Englewood, CO: Libraries Unlimited, Inc.

Salmon, G. (2000). *E-Moderating: The Key to Teaching and Learning Online.* London: Kogan Page.

Salomon, G. (1993). *Distributed Cognitions: Psychological and Educational Considerations.* New York: Cambridge U press.

Saussure, F.D. (1959). *Course in General Linguistics.* New York: McGraw-Hill.

Savery, J. R. (1998). Fostering ownership for learning with computer-supported collaborative writing in an undergraduate business communication course. In Bonk, C. J. and King, K. S. (Eds.), *Electronic Collaborators: Learner-Centered Technologies for Literacy, Apprenticeship, and Discourse.* Mahwah, NJ: Lawrence Erlbaum Associates.

Schank, R. C. (1997). *Virtual Learning.* New York: McGraw-Hill.

Schank, R. C. (1998). *Tell Me a Story: Narrative and Intelligence.* Evanston, IL: Northwestern University Press.

Schau, C. and Mattern, N. (1997). Use of map techniques in teaching applied statistics courses. *The American Statistician*, May.

Schillaci, A. and Culkin, J. M. (Eds.). (1970). *Films Deliver: Teaching Creatively With Film.* New York: Citation Press.

Schon, D. A. (1987). *Educating the Reflective Practitioner.* San Francisco, CA: Jossey-Bass Publishers.

Schwartz, D. L. (1999). The productive agency that drives collaborative learning. In Dillenbourg, P. (Ed.), *Collaborative Learning: Cognitive and Computational Approaches.* New York: Pergamon.

Senge, P. M. (2000). The academy as learning community: Contradiction in terms or realizable future? In A. F. Lucas and Associates (Eds.), *Leading Academic Change: Essential Roles for Department Chairs.* San Francisco, CA: Jossey-Bass.

Shaffer, D. W. and Resnick, M. (1999). "Thick" authenticity: New media and authentic learning. *Journal of Interactive Learning Research, 10*(2).

Shneiderman, B. (1987). *Designing the User Interface.* Reading, MA: Addison-Wesley Publishing Company.

Shneiderman, B. (1993). *Education By Engagement and Construction: Experiences in the AT&T Teaching Theater.* Keynote for ED-MEDIA 93, June, Orlando, FL AACE.

Shneiderman, B. (1995). Looking for the bright side of user interface agents. *ACM Interactions, 2*(1), 13-15.

Shneiderman, B. (1997). Direct manipulation for comprehensible, predictable and controllable user interfaces proc. *ACM International Workshop on Intelligent User Interfaces '97*, 33-39. New York: ACM.

Shneiderman, B. (1998). Codex, memex, genex: The pursuit of transformational technologies. *International Journal of Human-Computer Interaction, 10*(2), 87-106.

Shneiderman, B. (1999). Universal usability: Pushing human-computer interaction research to empower every citizen. Position Paper for *National Science Foundation & European Commission Meeting on Human-Computer Interaction Research Agenda*, June 1-4, Toulouse, France. To be published in book form.

Shneiderman, B. (1999b). Supporting creativity with advanced information-abundant user interfaces. Position Paper for *National Science Foundation & European Commission Meeting on Human-Computer Interaction Research Agenda*, June 1-4, Toulouse, France. To be published in book form.

Sholle, D. and Denski, S. (1994). *Media Education and the (Re)Production of Culture*. Westport, CT: Bergin & Garvey.

Sime, M. E. and Coombs, M. J. (1983). *Designing for Human-Computer Communication*. New York: Academic Press.

Skinner, B. F. (1964). *Technology of Teaching*. New York: Appleton-Century-Crofts.

Slavin, R. E. (1983). *Cooperative Learning*. New York: Longman.

Smith, J. and Marsiske, M. (1994). *Abilities and Competencies in Adulthood: Life-Span Perspectives on Workplace Skills*. National Center on Adult Literacy, October (ERIC: ED 377 339).

Sontag, S. (1977). *On Photography*. New York: Dell Publishing.

Spector, J. M. (1999). Teachers as designers of collaborative distance learning. *Proceedings of SITE 99*. Charlottesville, VA: AACE.

Sumner, T. and Taylor, J. (2000). Media integration through meta-learning environments. In Eisenstadt, M. and Vincent, T. (Eds.), *The Knowledge Web: Learning and Collaborating on the Net*. London: Kogan Page.

Tapscott, D. (1996). *The Digital Economy*. New York: McGraw-Hill.

The Institute for Higher Education Policy. (1999). *What's the Difference? A Review of Contemporary Research on the Effectiveness of Distance Learning in Higher Education*. Washington DC: The Institute for Higher Education Policy.

Thorson, M. K. (1999). *Campus-Free College Degrees: Thorson's Guide to Accredited Distance Learning Degree Programs*. Tulsa, OK: Thorson Guides.

Turkle, S. (1995). *Life on the Screen: Identity in the Age of the Internet*. New York: Simon & Schuster.

University Continuing Education Association. (1999). *Peterson's Independent Study Catalog*. Washington DC: University Continuing Education Association.

University Continuing Education Association. (2000). *Peterson's Guide to Distance Learning Courses*. Washington DC: University Continuing Education Association.

Vygotsky, L. (1997). *Thought and Language*. Cambridge, MA: The MIT Press.

Weiser, M. and Morrison, J. (1998). Project memory: Information management for project teams. *Journal of Management Information Systems*, Spring.

Weiss, G. and Dillenbourg, P. (1999). What is "multi" in multi-agent learning? In Dillenbourg, P. (Ed.), *Collaborative Learning: Cognitive and Computational Approaches*. New York: Pergamon.

Westland, J. C. (1994). Cinema theory, video games, and multimedia production. In Reisman, S. (Ed), *Multimedia Computing: Preparing for the 21st Century*. Hershey, PA: Idea Group Publishing.

Whatley, M. H. (1990). The picture of health: How textbook photographs construct health. In Ellsworth, E. and Whatley, M. H. (Eds.), *The Ideology of Images in Educational Media*. New York: Teachers College Press.

White, M. and Epston, D. (1990). *Narrative Means to Therapeutic Ends*. New York: W.W. Norton & Company.

Williams, R. L. (1972). *Documentary Photography*. New York: Time-Life Books.

Wolf, M. A. (1993). Learning: Meeting the challenges of older adulthood. Paper presented at the *Annual Meeting of the American Association for Adult and Continuing Education*, November 19 (ERIC ED 366 748).

Wollen, P. (1970). *Signs and Meaning in the Cinema*. Bloomington, IN: Indiana University Press.

Wooffitt, R. Fraser, N. M., Gilbert, N. and McGlashan, S. (1997). *Humans, Computers and Wizards*. New York: Routledge.

Worth, S. (1981). *Studying Visual Communication*. Philadelphia, PA: University of Pennsylvania Press.

Appendix A: Research Methodology

The original research data presented in this book are from two separate efforts: a Fall 2000 research involving the collection of both national data from higher education institutions, and a Winter 2001 survey of students enrolled in distance learning courses.

2000 AND 2001 SURVEYS

Some academics writing about research methods suggest that studies may benefit from a combined quantitative and qualitative approach (Campbell & Stanley, 1966; Gorden, 1975). The benefits are that data sets can be compared for consistency, and the interviews can allow some insight into the causal processes, while the surveys can provide indication of the prevalence of the phenomenon. Consequently, the methodology for the 2000 study consists of two parts: a quantitative questionnaire sent to administrators of distance learning programs and qualitative interviews of representatives at a sample of institutions.

For the purposes of this study, distance learning format courses are defined as having at least a 50% reduction in seat time through use of any technology or medium. Three primary lists of distance learning providers were used for the quantitative part of the study: *Peterson's Guide to Distance Learning Programs* (UCEA, 2000), *Peterson's Independent Study Catalog* (UCEA, 1999) and *Campus-Free College Degrees: Thorson's Guide to Accredited Distance Learning Degree Programs* (Thorsen, 1999). The lists were combined and duplicates eliminated.

In the first mailing, 1,114 emailed surveys were sent out. However, only 623 of the email addresses were sent directly to individuals; the rest went to generic department or university information addresses. Of those emailed surveys, 295 were returned non-deliverable. Five institutions responded that they did not have distance learning courses as defined in the study; four others declined to participate for various reasons. The final number of completed surveys returned was 176. If the returned emails are excluded, this constitutes a response rate of 21.5%. As this original list was quite comprehensive, including virtually all the higher education institutions in America using distance learning to any significant degree, this response is quite good. Additionally, in comparison to other national surveys of distance learning that are based on smaller samples (such as the ITC, NEA, and Primary Research Group surveys), this response is deemed sufficient.

In addition to this quantitative sample, representatives from 17 institutions were interviewed. The decision on choice of institutions for interviews was made based on two seemingly contradictory factors: first, to get a cross-section of types of institutions (two-year, four-year, rural, urban, research one, comprehensive, HBCU, religious, distance learning only, proprietary), and second, uniqueness in terms of being leaders in the field. Additionally, the choice of institutions to study through interviews followed the return of the survey in order to identify representative institutions and further probe issues identified through the quantitative data.

The survey for the quantitative part of the 2000 study was completed through the Internet. The benefits of this approach were that the Internet is widely used in higher education, is more convenient than regular mail, and allows for much faster response time (Heflich & Rice, 1999). The process was to send an introductory email that answered basic questions about the project, handled fears of confidentiality, and motivated people to participate. The email was linked to a URL (uniform resource locator) where a computer form was used to collect the data on a remote server. The surveys were coded so that responses were tracked (except for a few who thwarted the coding system). A follow-up email was sent two weeks later to remind those who had not responded to complete the survey.

For the interviews, potential subjects were contacted by email. The interview instrument itself consisted of main questions, probes, and follow-up questions. The questions focused on more in-depth probing of the primary and secondary questions of the study. The interviewees were asked to reserve 60 minutes for a single interview. Bernard (1988) suggests that in situations in which the researcher will have only one chance to interview, semi-structured interviewing is best. Accordingly, while there was a set of standard questions for each subject, it was intended that there would be flexibility to pursue other questions as indicated by the on-the-spot responses.

A survey of students who had completed distance learning format courses at Chapman University was conducted in early 2001. These included primarily professional development courses for K-12 teachers, and some individual undergraduate credit courses. The undergraduate courses were all videotape based, and were mostly "meet degree" requirements. The K-12 professional development courses were delivered online and through CD-ROM, and most were generally not required for a degree. A random sample of 659 students enrolling in distance learning courses over a five-year period (including computer-based, video, and correspondence format courses) were mailed surveys. 221 surveys were returned undeliverable due to expired addresses. 129 completed surveys were returned and processed. When the undeliverable surveys are excluded, this represents a 29.4%

response rate. The survey (Appendix J) centered on key questions brought forth from the research literature review. Additionally, because the respondents were taken from a sample of different types of distance learning delivery formats (videotape, CD-ROM, online), attitudes of students towards different distance learning methods were probed, and a section specifically for students taking computer-based courses was included to isolate those responses.

ANALYSIS

The quantitative data for the 2000 and 2001 surveys were tabulated and analyzed to describe current practices and approaches to distance learning in higher education. The population of higher education institutions in America offering distance learning credit courses is to a large extent known. However, because survey responses were depended upon for the data, inferential statistical methods were used to generalize about the whole population from the sample of respondents (Bernard, 1988; Healey, 1999). SPSS software was used to tabulate and analyze the data to identify significant descriptive patterns and relationships among the variables. For the 2001 survey, simple comparisons of totals by delivery method were primarily used. SPSS software was also used to analyze the data from the survey.

LIMITATIONS OF THE DATA

In regard to the 2000 study's limitations, it is important to note that those both surveyed and interviewed were primarily administrators. For the survey, 78.2% of respondents reported that they are classified as full-time administrators. While the interviews revealed that many full-time administrators had been faculty members at one time—this is likely also true of many completing the survey—the respondents' administrative backgrounds need to be recognized in evaluating the results.

The first limitation of the 2001 student survey is that the sample is rather small (129 respondents). Additionally, the respondents came from a pool of over 30 courses taken at different times, using different delivery methods and instructors. The use of different instructors certainly contributed to the different experiences reported by the students. Furthermore, as these are degree and continuing education courses aimed at adult students, the responses do not necessarily reflect attitudes of traditional undergraduate students or primary and secondary students. Nevertheless, this survey is useful particularly in understanding the learning styles and preferences of students choosing to enroll in distance learning courses.

Appendix B: 2000 Survey Quantitative Instrument

Distance Learning Survey

Please fill out the form below and **click the submit button at the end to complete**. Please note: for the purposes of this study, I define distance learning as at least 1/2 reduction in traditional face-to-face teacher-student contact using any delivery method (from mail to email, from videotape to videoconferencing). This survey is designed to be completed by the person most knowledgeable about your institution's distance education course offerings.

Your email address: ☐

Do you speak for the university as a whole in relationship to distance learning?

○ Yes

○ Maybe

○ No

1. Are you (check all that apply):

    ☐ full-time administrator

    ☐ part-time administrator

    ☐ tenured faculty

    ☐ part-time faculty

2. Highest degree your institution awards:

○ Associate

○ Bachelor's

○ Master's

○ Ph.D./Professional degree

3. Your institution is:

○ public

○ independent

4. Number of distance learning format (at least 1/2 reduction in face-to-face student-teacher contact) courses Summer 1999-Spring 2000?:

5. Number of distance learning format courses Summer 1998-Spring 1999?:

Implementation Motive

6. What do you see as the main reason(s) for your institution being involved in offering distance learning format programs? (mark all that apply)

☐ provide access to wider student population

☐ provide IT skills for students

☐ new source of revenue

☐ reduce expenses

☐ belief in teaching/learning advantages of DL

☐ desire to keep up with competition

☐ other, please specify: [   ]

7. Is providing greater access to student populations part of your institutional mission?

○ Yes

○ Maybe

○ No

8. Do you see providing information technology skills for students as a primary role for your institution?

○ Yes

○ Maybe

○ No

9. Is the pursuit of new sources of revenue for the institution through new program development a primary concern?

○ Yes

○ Maybe

○ No

10. Is the reduction of labor and facility costs a primary force in institutional planning?

○ Yes

○ Maybe

○ No

11. Do you believe that technology offers specific teaching/learning advantages?

○ Yes

○ Maybe

○ No

12. Is it necessary to offer distance learning format courses in order to keep up with competing institutions?

○ Yes

○ Maybe

○ No

13. Who initiated the use of distance learning courses at your institution?

- ○ top university administrator(s)
- ○ extended/continuing education deptartment
- ○ individual faculty
- ○ task force/committee
- ○ external agency
- ○ other, please specify: [   ]

14. Have the availability of funding sources and/or new state or federal government agencies encouraged the development of your distance learning programs?

- ○ Yes
- ○ Maybe
- ○ No

Administration/Management

15. In what administrative unit is the distance learning program now housed? : [   ]

16. Is this administrative unit budgetarily described as:

- ○ subsidized
- ○ partially subsidized

○ self-supporting

17. Is there a institutional distance learning plan?

○ Yes

○ No

○ Don't know

If so, who wrote it (position)? If a committee created it, what departments were represented?: ☐

18. Which of the following best describes the current economic status of your distance learning program:

○ large deficit

○ deficit

○ break even

○ profit

○ large profit

19. Does your institution have a different procedure for the academic approval of distance learning format courses?

○ Yes

○ No

If so, how is it different?: ☐

20. Does your institution use whole courses licensed from other educational institutions?

○ Yes

○ No

21. Are you using any kind of course brokering service to market distance learning courses to non-matriculated students?

○ Yes

○ No

22. Do you feel that the distance learning program is consistent with your institutional mission?

○ Yes

○ Maybe

○ No

23. Do you view distance learning in higher education as an administrative innovation?

○ Yes

○ Maybe

○ No

24. How are full-time faculty compensated for teaching distance learning format courses:

○ As regular load with normal enrollment limits

○ As regular load with no ceiling on enrollment

○ As regular load with additional pay after seat maximum

If so, what amount?: [   ]

○ With additional preparation time

If so how much?: [   ]

○ On an overload basis

If so, how calculated?: [   ]

○ With additional stipend

If so, what amount?: [   ]

○ As regular load for in-person class, per head for remote students

○ Other, please specify: [   ]

○ Do not use full-time faculty

25. Are faculty compensated differently for different types of technologies utilized (i.e., videotape versus Internet based)?

○ Yes (please specify difference): [   ]

○ No

26. Who determines which faculty/instructor will teach a distance learning course? (check all that apply)

☐ The academic department administration

☐ The distance education director

☐ The faculty/instructor

☐ Joint decision between faculty and administration

☐ Other, please specify: [   ]

27. What form of recognition is there for faculty/instructors teaching distance learning format courses? (check all that apply)

☐ Merit reimbursement

☐ Promotion

☐ Tenure

☐ None

☐ Other, please specify: [   ]

28. Please indicate the level of faculty/instructor training/professional development in relationship to distance learning courses.

○ Available

○ Recommended

○ Required

29. How is training provided? (check all that apply)

- [ ] in person at beginning
- [ ] in person on-going
- [ ] online
- [ ] print

30. When a full-time faculty member develops a distance learning course as part of either regular load, overload, or for a stipend, who owns the intellectual property rights?

- ○ faculty
- ○ institution
- ○ joint ownership
- ○ no policy on this issue at this time
- ○ other, please specify: [   ]
- ○ do not use full-time faculty

31. What is the percentage of faculty teaching distance learning courses that are classified as adjunct faculty?

- ○ 0-25%
- ○ 26-50%
- ○ 51-75%
- ○ 76%+

32. Approximately how many technical support staff per faculty/instructor?: ▢

33. How many instruction design experts available per faculty/instructor?: ▢

34. Are courses typically developed by faculty/instructors through:

○ teams

○ individuals

○ both

35. Do you generally require a proctored test for course completion?

○ always

○ sometimes

○ never

36. Have you done a cost/benefit analysis of distance learning?

○ yes

○ no

If so, what does it show?: ▢

Pedagogy

37. Which of the following technologies is used as the primary form of delivery for your distance learning courses?

○ Internet

○ CD-ROM

○ pre-packaged videotape (not live)

○ live video (one-way or two-way)

○ telephone (audio only)

○ print-based

○ other, please specify: [   ]

38. Do you view distance learning in higher education as a teaching/learning innovation?

○ yes

○ maybe

○ no

39. What evidence do you have that students learn effectively in distance learning courses? What does it show?: [   ]

Please rate your degree of agreement with the following statements in regard to your general institutional approach to distance learning courses.

40. Questions about how best to convert course material to distance learning format are of great importance.

○ strongly agree

○ agree

○ disagree

○ strongly disagree

41. Courses include significant interaction with other students.

○ strongly agree

○ agree

○ disagree

○ strongly disagree

42. Courses offer opportunity to collaborate with other students on projects.

○ strongly agree

○ agree

○ disagree

○ strongly disagree

43. Courses include simulations and/or case studies.

○ strongly agree

○ agree

○ disagree

○ strongly disagree

44. Great care is taken in understanding how students navigate through the course software.

○ strongly agree

○ agree

○ disagree

○ strongly disagree

45. Courses are like one-on-one tutoring with the faculty member, providing rich and prompt feedback to the students.

○ strongly agree

○ agree

○ disagree

○ strongly disagree

46. Taking a distance learning course at your institution is most like (choose one or more):

☐ reading a book

☐ watching TV

- ☐ watching a movie
- ☐ listening to the radio
- ☐ talking on the telephone
- ☐ writing letters

47. How long does it generally take to develop a new distance learning format course?

- ○ less than six months
- ○ 1 year
- ○ 2 years
- ○ More than 2 years

48. Which of the following most closely reflects the course development process?

- ○ existing course material automated by technology
- ○ existing course material automated, with some new material
- ○ all course material developed specifically for distance learning course

49. Is assessment of student learning similar or the same as in traditional courses?

- ○ Yes
- ○ No

50. Additional comments?: [   ]

I would be available for a follow-up short interview over the phone.

○ Yes

○ No

Thank you for taking the time to complete this survey. Please complete the following form if you would like to receive the results from this project.

Name: [   ]

Address1: [   ]

Address2: [   ]

City/State/ZIP: [   ]

[Submit Survey]

Appendix C: 2000 Qualitative Instrument

Qualitative Survey

General Demographic Information

1. Institution Name:

2. Interviewee:

3. Title:

Implementation Motive

4. What do you see as the main reasons for your institution being involved in offering distance learning format programs?

5. Describe the process of implementing the use of distance learning courses at your institution?

Administration/Management

6. Describe the administration and management of distance learning here.

7. What are the key aspects of the formal or informal university distance learning plan?

8. Describe how the academic oversight for distance learning courses is the same or different from traditional courses.

9. What has been your approach to collaborations with for-profit company services?

10. In what ways does distance learning fit the overall mission of your institution?

11. How is the use of faculty for distance learning courses the same or different from traditional courses?

Pedagogy

12. Describe the technologies you use in your distance learning courses and why you use them.

13. How do you address the issue of student interaction in your distance learning courses?

14. What approaches do you use to develop a sense of community in your distance learning courses?

15. How are case studies and simulations used in your distance learning courses?

16. What is your approach to designing computer user navigation schemes in your courses?

17. What is the development process for new distance learning format courses?

18. How is assessment of student learning different from traditional courses?

Appendix D: 2000 Survey Cover Email

11/2/2000

Attn: Educators Involved in Distance Learning (please forward within your institution if necessary)

The increased use of distance learning is one of the most controversial subjects in American higher education today. Yet clear and comprehensive information on why and how distance learning is being implemented in higher education is hard to find. By completing the linked survey you can contribute to a better understanding of distance learning in higher education, and receive the results of the study as well.

As part of a dissertation project at Claremont Graduate University, I am conducting research on the implementation and practice of distance learning in higher education. This research focuses on patterns of distance learning administration in higher education, including reasons for adoption and administrative practices.

If you do not feel qualified to complete this survey, please forward it to the person at your institution most knowledgeable about distance learning. If you do not believe that your institution is involved in distance learning of any kind, please reply to this email with that statement.

The survey is completed through a convenient computer form. Simply go to the following URL: http://www.webcom.com/exed.

Thank you.

Gary A. Berg
gberg99us@yahoo.com

Appendix E: 2000 Survey Email to Interview Subjects

Dear XXXX,

The increased use of distance learning is one of the most controversial subjects in American higher education today. Yet clear and comprehensive information on why and how distance learning is being implemented in higher education is hard to find.

As part of a dissertation project at Claremont Graduate University, I am conducting research on the implementation and practice of distance learning in higher education. This research focuses on patterns of distance learning administration in higher education, including reasons for adoption and administrative practices. I ask that you take the time for a short interview either in person or over the phone.

The interview should take approximately 60 minutes to complete. If you need additional information or would like to discuss the project, please call or email.

I thank you.

Sincerely,

Gary A. Berg
gberg99us@yahoo.com

Appendix F: 2000 Survey Online Statement of Informed Consent

Distance Learning Survey

ONLINE STATEMENT OF INFORMED CONSENT FOR RESEARCH PARTICIPATION

By clicking on "I AGREE" below, I hereby give my consent to participate as a subject in an investigation conducted by Gary A. Berg into distance learning practices in higher education as part of his dissertation research at Claremont Graduate University. I understand that: responses will be kept in strictest confidence, and the results of this project will be coded in such a way that my identity will not be attached in any way to the final data that is produced. In the event that I have any questions, I can contact the researcher, Gary A. Berg, at (714) 532-6049 or by email (gberg@chapman.edu). If you agree to participate, please double click here:

I agree

Appendix G: 2000 Survey In-Person Statement of Informed Consent

STATEMENT OF INFORMED CONSENT

I agree to participate voluntarily in a research study on distance learning in higher education conducted by Gary A. Berg as part of his dissertation project at Claremont Graduate University. The information that I provide will be used to conduct research, the purposes of which is to examine distance learning practices in higher education through the experiences and perspectives of the individuals participating in the study.

I understand that the interview includes questions concerning my beliefs, choices, and life events. The interview should take approximately 30-45 minutes. I may decline to answer any question and/or terminate the interview at any time without any reservation whatsoever. Should I have any questions or concerns about the research or about my participation, the interviewer will address them.

I understand that the interview will be tape recorded. If I myself choose to disclose my participation in this study or any part of the interview or if I authorize the investigators to do so, I will take responsibility for any risk associated with such disclosure. My name will used in publications resulting from this study only if I give my written permission. I will be given the option to disclose my participation in the study and have quotes attributed to me after I have completed the interview. I can also choose to read the transcript of my interview, and then express my wishes on a response form.

I understand that I will receive no payment for my participation in this study.

I have read the consent form and fully understand it. All my questions have been answered. I agree to take part in the study.

_____   Gary Berg_____
Printed name of Participant   Name of interviewer

_____   _____
Signature of Participant   Signature of interviewer

_____
Date

Please mail to:

Gary A. Berg
gberg99us@yahoo.com

Appendix H: 2000 Survey Post-Interview Form

Response Form
Post-Interview

1. I give permission to be identified by name when quotes from my interview are used:

( ) Yes
( ) No

2. I give permission to be identified by name in the complete list of interviewees when you publish the results of the study:

( ) Yes
( ) No

_____
Printed Name

_____   _____
Signature   Date

Appendix I: 2000 Survey List of Participating Institutions

Institutions Responding to Survey	Carnegie Classification	State
American Military University	Specialized Institutions	VA
Assemblies of God Theological Seminary	Specialized Institutions	MO
Atlantic Cape Community College	Associate's Colleges	NJ
Bakersfield College	Associate's Colleges	CA
Beaufort County Community College	Associate's Colleges	NC
Boston University	Doctoral/Research U-Ext	MA
Bradley University	Master's Colleges & U I	IL
Burlington County College	Associate's Colleges	NJ
Butte College	Associate's Colleges	CA
Central Piedmont Community College	Associate's Colleges	NE
Central Virginia Community College	Associate's Colleges	VA
Central Wyoming College	Associate's Colleges	WY
Chadron State College	Master's Colleges & UI	NE
Chicago State University	Master's Colleges & UI	IL
City Colleges of Chicago, Washington	Associate's Colleges	IL
Clackmas Community College	Associate's Colleges	OR
Community Hospital of Roanoke Valley	Specialized Institutions	VA
Connecticut College	Baccalaureate Coll.-Lib	CT
Cossatot Technical College	Associate's Colleges	AR
California State University, Chico	Master's Colleges & UI	CA
California State University, San Diego	Master's Colleges & UI	CA
California State University, San Marcos	Master's Colleges & UI	CA
California State University, Sonoma	Master's Colleges & UI	CA
Dallas Baptist University	Master's Colleges & UI	TX
Dallas Theological Seminary	Specialized Institutions	TX
Des Moines Area Community College	Associate's Colleges	IA
Dodge City Community College	Associate's Colleges	KS
Edison State Community College	Master's Colleges & UII	NJ

EDUKAN (consortium)	Associate's Colleges	KS
Emporia State University	Master's Colleges & UI	KS
Florida Agricultural & Mechanical U	Master's Colleges & UI	FL
Florida State University	Doctoral/Research U-ext	FL
Friends University	Master's Colleges & UI	KS
Genesee Community College	Associate's Colleges	NY
Georgia State University	Doctoral/Research U-ext	GA
Greenville Technical College	Associate's Colleges	SC
Horry-Georgetown Technical College	Associate's Colleges	SC
Hudson Valley Community College	Associate's Colleges	NY
Iowa Western Community College	Associate's Colleges	IA
Jacksonville State University	Master's Colleges & UI	AL
James Madison University	Master's Colleges & UI	VA
Johnson Bible College	Specialized Institutions	TN
Keiser College	Associate's Colleges	FL
Kellogg Community College	Associate's Colleges	MI
Kennesaw State University	Master's Colleges & UI	GA
Kirksville College of Osteopathic Medicine	Specialized Institutions	MO
Labette Community College	Associate's Colleges	KS
Lackawanna Junior College	Associate's Colleges	PA
Lake Sumter Community College	Associate's Colleges	FL
Lakeland Community College	Associate's Colleges	OH
Lansing Community College	Associate's Colleges	MI
Lehigh University	Doctoral/Research-Ext	PA
Louisiana State University & Agriculture	Doctoral/Research-Ext	LA
Mansfield University of Pennsylvania	Master's Colleges & UI	PA
Marian College	Baccalaureate Coll.—G	IN
Michigan State University	Doctoral/Research-Ext	MI
Midland College	Associate's Colleges	TX
Mid-State Technical College	Associate's Colleges	WI
Minnesota West Community & Technical	Associate's Colleges	MN
Mira Costa College	Associate's Colleges	CA
Mississippi County Community College	Associate's Colleges	AR
Missouri Western State College	Baccalaureate Coll.—G	MO
Mount San Antonio College	Associate's Colleges	CA
Naropa University	Specialized Institutions	CO
Newman University	Master's Colleges & UI	KS
North Arkansas College	Associate's Colleges	AR
North Central University	Specialized Institutions	MN
North Dakota State College of Science	Associate's Colleges	ND

North Iowa Area Community College	Associate's Colleges	IA
Northcentral Technical College	Associate's Colleges	WI
Northeastern University	Doctoral/Research-Ext	MA
Northern Michigan University	Master's Colleges & UI	MI
Nyack College	Master's Colleges & UI	NY
Oklahoma State University	Doctoral/Research-Ext	OK
Open Learning Fire Service Program	Baccalaureate Colleges–G	MD
Palo Alto College	Associate's Colleges	TX
Palomar College	Associate's Colleges	CA
Pellissippi State Technical Comm College	Associate's Colleges	TN
Piedmont College	Master's Colleges & UI	GA
Plattsburgh State University of New York	Master's Colleges & UI	NY
Pratt Community College	Associate's Colleges	KS
Prescott College	Master's Colleges & UII	AZ
Red Rocks Community College	Associate's Colleges	CO
Robert Morris College	Specialized Institutions	IL
Saddleback College	Associate's Colleges	CA
Salt Lake Community College	Associate's Colleges	UT
San Diego City College	Associate's Colleges	CA
Sandhills Community College	Associate's Colleges	NC
Santa Fe Community College	Associate's Colleges	FL
Saybrook Institute	Specialized Institutions	CA
Siena Heights University	Master's Colleges & UI	MI
Southeast Community College—Beatrice	Associate's Colleges	NE
Southern Oregon University	Master's Colleges & UI	OR
Southern Utah University	Master's Colleges & UII	UT
Southwest State University	Baccalaureate Coll.–G	MN
Southwestern Community College	Associate's Colleges	SC
St. Cloud State University	Master's Colleges & UI	MN
St. Petersburg Junior College	Associate's Colleges	FL
State University of NJ, New Brunswick	Doctoral/Research-Ext	NJ
State University of NY—Binghamton	Doctoral/Research-Ext	NY
Stevens Institute of Technology	Doctoral/Research-Int	NJ
Suffolk County Community College	Associate's Colleges	NY
Sussex County Community College	Associate's Colleges	NJ
Taylor University	Baccalaureate Coll.—G	IN
Texas A&M University—Commerce	Doctoral/Research-Ext	TX
Texas Wesleyan University	Master's Colleges & UII	TX
Thomas Edison State College	Master's Colleges & UII	NJ
Tidewater Community College	Associate's Colleges	VA

Tiffin University	Specialized Institutions	OH
Treasure Valley Community College	Associate's Colleges	OR
Troy State University—Dothan	Master's Colleges & UI	AL
University of Alabama	Doctoral/Research-Ext	AL
University of Alaska, Southeast	Master's Colleges & UI	AK
University of Arkansas	Doctoral/Research-Ext	AR
University of Central Florida	Doctoral/Research-Int	FL
University of Charleston	Baccalaureate Coll.—G	WV
University of Colorado—Boulder	Doctoral/Research-Ext	CO
University of Illinois at Urbana—Champaign	Doctoral/Research-Ext	IL
University of Iowa	Doctoral/Research-Ext	IA
University of Louisville	Doctoral/Research-Ext	KY
University of Maine—Fort Kent	Baccalaureate Coll.—G	ME
University of Maine—Machias	Baccalaureate Coll.—G	ME
University of Nebraska—Lincoln	Doctoral/Research-Ext	NE
University of Nevada, Reno	Doctoral/Research-Ext	NV
University of North Dakota	Doctoral/Research-Int	ND
University of Oregon	Doctoral/Research-Ext	OR
University of Pittsburgh	Doctoral/Research-Ext	PA
University of Saint Francis	Master's Colleges & UI	IL
University of Sioux Falls	Master's Colleges & UII	SD
University of South Dakota	Doctoral/Research-Int	SD
University of Southern Mississippi	Doctoral/Research-Ext	MS
University of Tennessee, Knoxville	Doctoral/Research-Ext	TN
University of Vermont	Doctoral/Research-Ext	VT
University of Virginia	Doctoral/Research-Ext	VA
University of Wisconsin—Madison	Doctoral/Research-Ext	WI
Utah Valley State College	Associate's Colleges	UT
Virginia Commonwealth University	Doctoral/Research-Ext	VA
West Texas A & M University	Master's Colleges & UI	TX
West Virginia Northern Community College	Associate's Colleges	WV
West Virginia University	Doctoral/Research-Ext	WV
Western Seminary	Specialized Institutions	OR
Wytheville Community College	Associate's Colleges	VA
York College of Pennsylvania	Master's Colleges & UII	PA
York Technical College	Associate's Colleges	SC

32 anonymous institutions (decided to not use survey coding and remain anonymous)

The 2000 Carnegie Classification includes all colleges and universities in the United States that are degree-granting and accredited by an agency recognized by the U.S. Secretary of Education. The 2000 edition classifies institutions based on their degree-granting activities from 1995-96 through 1997-98.

Doctorate-Granting Institutions

Doctoral/Research Universities—Extensive: These institutions typically offer a wide range of baccalaureate programs, and they are committed to graduate education through the doctorate. During the period studied, they awarded 50 or more doctoral degrees[1] per year across at least 15 disciplines.[3]

Doctoral/Research Universities—Intensive: These institutions typically offer a wide range of baccalaureate programs, and they are committed to graduate education through the doctorate. During the period studied, they awarded at least ten doctoral degrees[1] per year across three or more disciplines,[2] or at least 20 doctoral degrees per year overall.

Master's Colleges and Universities

Master's Colleges and Universities I: These institutions typically offer a wide range of baccalaureate programs, and they are committed to graduate education through the master's. During the period studied, they awarded 40 or more master's degrees per year across three or more disciplines.[2]

Master's Colleges and Universities II: These institutions typically offer a wide range of baccalaureate programs, and they are committed to graduate education through the master's degree. During the period studied, they awarded 20 or more master's degrees per year.

Baccalaureate Colleges

Baccalaureate Colleges—Liberal Arts: These institutions are primarily undergraduate colleges with major emphasis on baccalaureate programs. During the period studied, they awarded at least half of their baccalaureate degrees in liberal arts fields.[3]

Baccalaureate Colleges—General: These institutions are primarily undergraduate colleges with major emphasis on baccalaureate programs. During the period studied, they awarded less than half of their baccalaureate degrees in liberal arts fields.

Baccalaureate/Associate's Colleges: These institutions are undergraduate colleges where the majority of conferrals are at the subbaccalaureate level (associate's degrees and certificates). During the period studied, bachelor's degrees accounted for at least ten percent but less than half of all undergraduate awards.

Associate's Colleges

These institutions offer associate's degree and certificate programs but, with few exceptions award no baccalaureate degrees.[4] This group includes institutions where, during the period studied, bachelor's degrees represented less than 10 percent of all undergraduate awards.

Specialized Institutions

> These institutions offer degrees ranging from the bachelor's to the doctorate, and typically award a majority of degrees in a single field. The list includes only institutions that are listed as separate campuses in the *Higher Education Directory*. Specialized institutions include:
>
> > Theological seminaries and other specialized faith-related institutions: These institutions primarily offer religious instruction or train members of the clergy.
> >
> > Medical schools and medical centers: These institutions award most of their professional degrees in medicine. In some instances, they include other health professions programs, such as dentistry, pharmacy, or nursing.
> >
> > Other separate health profession schools: These institutions award most of their degrees in such fields as chiropractic, nursing, pharmacy, or podiatry.
> >
> > Schools of engineering and technology: These institutions award most of their bachelor's or graduate degrees in technical fields of study.
> >
> > Schools of business and management: These institutions award most of their bachelor's or graduate degrees in business or business-related programs.
> >
> > Schools of art, music, and design: These institutions award most of their bachelor's or graduate degrees in art, music, design, architecture, or some combination of such fields.
> >
> > Schools of law: These institutions award most of their degrees in law.
> >
> > Teachers colleges: These institutions award most of their bachelor's or graduate degrees in education or education-related fields.
> >
> > Other specialized institutions: Institutions in this category include graduate centers, maritime academies, military institutes, and institutions that do not fit any other classification category.

The above definition of classifications quoted directly from Carnegie Foundation for the Advancement of Teaching. (2000). *The Carnegie Classification of Institutions of Higher Education, 2000 Edition.* Electronic data file.

Institutions Participating in Interviews

Bakersfield College, Greg Chamberlain, Dean of Learning Resources

Bellevue Community College, Thornton Perry, Director of Distance Education

Boston University, Elizabeth Spencer-Dawes, Manager, Distance Learning

Chapman University, Don Cardinal, Faculty

California State University, Dominguez Hills, Warren Ashley, Director, Mediated Instruction and Distance Learning

Florida State University, Carole Hayes, Coordinator External Relations

Foothill Community College, Vivian Sinou, Dean of Distance & Mediated Learning

Georgia State University, Jacquelynn Sharpe, Coordinator, Division of Distance and Distributed Learning.

Nassau Community College, Arthur Friedman, Coordinator, College of the Air

Saybrook Institute, Kathy Wiebe, Admissions Coordinator

St. Cloud State University, John Burgeson, Dean of Continuing Education

Texas Wesleyan University, Joy Edwards, Director of Graduate Studies

Western Seminary, Jon Raibley, Assistant Director of Lifelong Learning Center

Vice-President, anonymous large, independent, Eastern U. S. doctoral degree-granting institution

Public Relations Director, anonymous large, public, Southern U. S. doctoral degree-granting institution

Faculty and Program Director, anonymous Historically Black College and University.

Program Manager, anonymous large, independent, Western doctoral degree-granting institution.

Appendix J: 2001 Survey Letter to Subjects

September 1, 2001

Dear Student,

In order to better serve Chapman University students enrolled in our distance learning courses, we are surveying those who have taken courses in alternative formats including videotape, correspondence, and computer-based delivery methods.

Please take a moment to complete the enclosed survey and return it in the business reply envelope. Your participation is greatly appreciated.

Sincerely,

Gary A. Berg
Director of Extended Education and Distance Learning
gberg99us@yahoo.com

Appendix K: 2001 Quantitative Instrument

Distance Learning Survey

1. Delivery format:
___ computer-based
___ videotape
___ correspondence
___ other _____

Please indicate your degree of agreement with the following statements:

2. I learned as much or more in this distance learning course as in an average traditional face-to-face course.
___ strongly agree   ___ agree   ___ disagree   ___ strongly disagree

3. The quality of the interaction with the instructor was the same or better as in a traditional face-to-face course.
___ strongly agree   ___ agree   ___ disagree   ___ strongly disagree

4. The quality of the interaction with other students in the course was the same or better as in a traditional face-to-face course.
___ strongly agree   ___ agree   ___ disagree   ___ strongly disagree

5. Critical thinking skills were utilized and developed in this course.
___ strongly agree   ___ agree   ___ disagree   ___ strongly disagree

6. I primarily learned in this course through memorization and repetition.
___ strongly agree   ___ agree   ___ disagree   ___ strongly disagree

7. The media used in this course added significantly to the learning experience.
___ strongly agree   ___ agree   ___ disagree   ___ strongly disagree

8. I would prefer active interaction with the course material, instructor, and other students over recorded lectures and prepared materials.
___ strongly agree   ___ agree   ___ disagree   ___ strongly disagree

9. It is important that courses are customized to meet my specific learning needs and to adjust to my learning style.
___ strongly agree   ___ agree   ___ disagree   ___ strongly disagree

10. I like to have group projects and other opportunities to learn in group situations.
___ strongly agree   ___ agree   ___ disagree   ___ strongly disagree

11. I would prefer to communicate with other students around interdependent tasks, rather than open socializing opportunities at a distance.
___ strongly agree   ___ agree   ___ disagree   ___ strongly disagree

12. Video is useful in showing or demonstrating content.
___ strongly agree   ___ agree   ___ disagree   ___ strongly disagree

13. Clarity of voice-over narration is more important than realistic sound levels.
___ strongly agree   ___ agree   ___ disagree   ___ strongly disagree

14. I would prefer some background music for distance learning courses.
___ strongly agree   ___ agree   ___ disagree   ___ strongly disagree

15. Photographs are evidence of the truthfulness of argued points in the subject matter.
___ strongly agree   ___ agree   ___ disagree   ___ strongly disagree

16. When course content is put into the form of a story it is easier to understand.
___ strongly agree   ___ agree   ___ disagree   ___ strongly disagree

17. I would like the distance learning format course to be structured like a tutorial with one-on-one contact with a tutor.
___ strongly agree   ___ agree   ___ disagree   ___ strongly disagree

18. I would like my distant learning to be well integrated with my work/job.
___ strongly agree   ___ agree   ___ disagree   ___ strongly disagree

19. I am very much aware of the point of view of the author(s) in representations made with various course media.
___ strongly agree   ___ agree   ___ disagree   ___ strongly disagree

Answer the following only if you took a computer-based course.

20. I like maximum control over how to navigate through the course software.
___ strongly agree   ___ agree   ___ disagree   ___ strongly disagree

21. The use of case studies and computer simulations increases learning for me.
___ strongly agree   ___ agree   ___ disagree   ___ strongly disagree

22. I would like to have the computer serve as a learning assistant or agent in distance learning courses.
___ strongly agree   ___ agree   ___ disagree   ___ strongly disagree

23. I would like to use computers for creative and visualization purposes in distance learning courses.
___ strongly agree   ___ agree   ___ disagree   ___ strongly disagree

Further Comments:

About the Author

Gary A. Berg, Ph.D., is author of *Why Distance Learning?* and over twenty articles on higher education and the use of technology in education. He has worked in higher education for many years as an administrator and is currently Director of Extended Education and Distance Learning at California State University, Channel Islands. Dr. Berg has developed many distance learning format courses and programs, been interviewed for numerous national publications, and consulted by educational and government organizations on the use of distance learning.

Index

A

abstract knowledge 180
Academic Literature on Film 112
accommodation 15
activity theory 50
adaptive agents 35
Adult Learning Theory 19
American still photography 122
animation 67
artificial intelligence 28
assimilation 15
audience expectations 139
author bias 124
automated self-tests 33
automated tutors 28

B

behaviorism 14
Brechts Bertolt 125
bricolage 90
British Open University method 31, 55
Bunuel, Luis 125, 170

C

camera angles 140
Capability Learning Model 25
Career Development Model (CDM) 25
case-based reasoning (CBR) 184
case studies 175, 184
chat rooms 41
Chautauqua Movement 10
chiaroscuro lighting 162
cognitive amplification 91
Cognitive Film Theory 131

collaborative learning 50
command language systems 64
communications theory 115
composition 160
computer agents 28
computer applications 134
computer environments 49
computer interface 179
computer interface design 63
computer simulations 175, 184
Computer-Assisted Instruction (CAI) 14
computer-based courses 101
computer-based educational environments 118, 187
computer-based learning 1
computer-based teaching methods 9
computer-based techniques 34
computer-based training (CBT) 4, 15
Computer-Mediated Conferencing 54
Computer-Supported Collaborative Learning (CSCL) 50
concept maps 86, 197
constructivist theory 15, 191
cooperative learning 9, 47
cooperative learning methods 17
Cooperative Learning Theory 16
critical thinking 86
cultural-historical psychology/activity theory (CHAT) 50
customization 28

D

direct manipulation systems 64
distance learning 34, 45, 50, 104
distance learning courses 1

distance learning format courses 101
documentary film 122
dramatic structure 135
dreaming 163

E

Edison, Thomas 10
editing 142
educational applications 175
educational environments 3
educational films 11
educational literature 179
educational radio 12
educational settings 179
educational technology 9
educational television 12
educational theorists 4
Eisenstein Sergei 144
empirical reality 129
equilibrium 15

F

faculty 93
face-to-face courses 101
film editing 142
film narrative 104
film style 113
film technology 10
film theory 104, 159
filmmakers 1
flow theory 79
folk psychology 181
foundational knowledge 48

G

genre 139
gestalt 62
Gordard, Jean-Luc 125
GOMS model 63
graphic user interfaces (GUI) 64
grounding 50
Group Learning 39, 195
Group Learning Theories 47
group projects 41
group work 44

H

heroes in film 169
high concept 136
history of educational technology 3
history of film 110
holistic-dynamic theory 20
human factors 61
human-computer interaction (HCI) 5, 60, 114
human-to-human tutorial communication 33
hypermedia 78
hypertext 65
hypertext environments 183

I

individual needs 32
information filtering 36
instructional films 112
instructional television (ITV) 12
intelligent agents 35, 202
intelligent computer-assisted instruction (ICAI) 14
intelligent tutoring systems (ITS) 14
intelligent tutors 35, 202
interactivity 70
interdependent tasks 52
interface design 63
interface metaphors 66

K

knowledge media 127

L

leadership 52
learning assistants 28
learning community 46
learning devices 202
learning environments 179
learning guide 32
learning society 21
learning styles 18
Learning Theory 9, 14
literacy 90

M

man-machine interface (MMI) 60
materialism 152
meaning making 179
media mix 127
media theory 104, 115
media viewing 163
medium 187
mentorships 31
metaphor 66
microworlds 90
mindtools 86, 91
montage theory 145
multimedia authoring program 134
multiple intelligences 18

N

narrative structure 136, 150, 175, 183
National Defense Education Act (NDEA) 13
navigation issues 80

O

on-the-job training 23
one-on-one tutoring 30

P

personal learning manager 34
personal stories 179
personal story construction 175
phantasmagoria 110
phenomenology of film 121
phonograph 12
Piaget, Jean 15
point of view 124, 159
power of words 3
predictive text generation 37
premise 136
profiling 28
project management 53
psychoanalysis and film 169
Public Broadcast System (PBS) 13

R

radio stations 12
reflection in action 22

Residential Conferences (RCs) 31

S

screening students 33
self-actualization 20
self-regulated learning 18
semantic-level grounding 50
semiology 128
simulations 175, 184
Skinner, B.F. 17
social communication 53
social learning theory 47
sound 155
specialized institution 32
still photography 118
story format 175
story slides 111
story telling 185
structural/materialism 152
student collaboration 40
student-centered system 48
subjectivity 159
surrealist movement 125

T

task analysis 68
teacher-learner relationship 31
teaching issues 93
text filtering 36
theme and plot 138
threaded conversations 53
tool 187
traditional classroom 1
tutor-student interaction 32
tutorial method 28

U

usability 62

V

verbalism 10
video courses 74
videotape courses 33
virtual learning community 46
virtual teams 51
Visual Instruction Movement 10
visual metaphor 148

visual perception 66
visualization 37, 201
voice-over narration 56, 198
Vygotsky, Lev 16

W

workplace learning 24

Knowledge Management and Business Model Innovation

Yogesh Malhotra, PhD, Syracuse University, USA

ISBN: 1-878289-98-5; Copyright: 2001
Pages: 464 (h/c); Price: US $149.95; Available: Now!

Knowledge Media and Healthcare: Opportunities and Challenges

Rolf Grutter, PhD, University of St. Gallen, Switzerland

ISBN: 1-930708-13-0; eISBN: 1-59140-006-6
Copyright: 2002; Pages: 296 (h/c); Price: US $74.95
Available: Now!

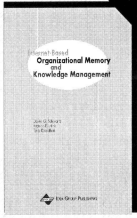

NEW Knowledge Management books from IGI!

Internet-Based Organizational Memory and Knowledge Management

David G. Schwartz, PhD, Bar-Ilan University, Israel
Monica Divitini, PhD, Norwegian University of Science and Technology
Terje Brasethvik, PhD, Norwegian University of Science and Technology

ISBN: 1-878289-82-9; eISBN:1-930708-63-7
Copyright: 2000; Pages: 269 (s/c); Price: US $119.95
Available: Now!

Knowledge Management and Virtual Organizations

Yogesh Malhotra, PhD, Syracuse University, USA

ISBN: 1-878289-73-X; eISBN:1-930708-65-3
Copyright: 2000; Pages: 450 (h/c)
Price: US $149.95; Available: Now!

Visit the Idea Group Online bookstore at www.idea-group.com for detailed information on these titles!

Recommend these KM books to your library!

NEW from Idea Group Publishing

- **Digital Bridges: Developing Countries in the Knowledge Economy**, John Senyo Afele/ ISBN:1-59140-039-2; eISBN 1-59140-067-8, © 2003
- **Integrated Document Management : Systems for Exploiting Enterprise Knowledge**, Len Asprey and Michael Middleton/ ISBN: 1-59140-055-4; eISBN 1-59140-068-6, © 2003
- **Critical Reflections on Information Systems: A Systemic Approach**, Jeimy Cano/ ISBN: 1-59140-040-6; eISBN 1-59140-069-4, © 2003
- **Web-Enabled Systems Integration: Practices and Challenges**, Ajantha Dahanayake and Waltraud Gerhardt ISBN: 1-59140-041-4; eISBN 1-59140-070-8, © 2003
- **Public Information Technology: Policy and Management Issues**, G. David Garson/ ISBN: 1-59140-060-0; eISBN 1-59140-071-6, © 2003
- **Knowledge and Information Technology Management: Human and Social Perspectives/** Angappa Gunasekaran, Omar Khalil and Syed Mahbubur Rahman/ ISBN: 1-59140-032-5; eISBN 1-59140-072-4, © 2003
- **Building Knowledge Economies: Opportunities and Challenges**, Liaquat Hossain and Virginia Gibson/ ISBN: 1-59140-059-7; eISBN 1-59140-073-2, © 2003
- **Knowledge and Business Process Management**, Vlatka Hlupic/ISBN: 1-59140-036-8; eISBN 1-59140-074-0, © 2003
- **IT-Based Management: Challenges and Solutions**, Luiz Antonio Joia/ISBN: 1-59140-033-3; eISBN 1-59140-075-9, © 2003
- **Geographic Information Systems and Health Applications**, Omar Khan/ ISBN: 1-59140-042-2; eISBN 1-59140-076-7, © 2003
- **The Economic and Social Impacts of E-Commerce**, Sam Lubbe /ISBN: 1-59140-043-0; eISBN 1-59140-077-5, © 2003
- **Computational Intelligence in Control**, Masoud Mohammadian, Ruhul Amin Sarker and Xin Yao/ISBN: 1-59140-037-6; eISBN 1-59140-079-1, © 2003
- **Decision-Making Support Systems: Achievements and Challenges for the New Decade**, M.C. Manuel Mora and Guisseppi Forgionne/ISBN: 1-59140-045-7; eISBN 1-59140-080-5, © 2003
- **Architectural Issues of Web-Enabled Electronic Business**, Shi Nan Si and V.K. Murthy/ ISBN: 1-59140-049-X; eISBN 1-59140-081-3, © 2002
- **Adaptive Evolutionary Information Systems**, Nandish V. Patel/ISBN: 1-59140-034-1; eISBN 1-59140-082-1, © 2003
- **Managing Data Mining Technologies in Organizations: Techniques and Applications**, Parag Pendharkar ISBN: 1-59140-057-0; eISBN 1-59140-083-X, © 2002
- **Intelligent Agent Software Engineering**, Valentina Plekhanova & Stefan Wermter/ ISBN: 1-59140-046-5; eISBN 1-59140-084-8, © 2003
- **Advances in Software Maintenance Management: Technologies and Solutions**, Macario Polo, Mario Piattini and Francisco Ruiz/ ISBN: 1-59140-047-3; eISBN 1-59140-085-6, © 2003
- **Multidimensional Databases: Problems and Solutions**, Maurizio Rafanelli/ISBN: 1-59140-053-8; eISBN 1-59140-086-4, © 2003
- **Information Technology Enabled Global Customer Service**, Tapio Reponen/ISBN: 1-59140-048-1; eISBN 1-59140-087-2, © 2003
- **Creating Business Value with Information Technology: Challenges and Solutions**, Namchul Shin/ISBN: 1-59140-038-4; eISBN 1-59140-088-0, © 2002
- **Advances in Mobile Commerce Technologies**, Ee-Peng Lim and Keng Siau/ ISBN: 1-59140-052-X; eISBN 1-59140-089-9, © 2003
- **Mobile Commerce: Technology, Theory and Applications**, Brian Mennecke and Troy Strader/ ISBN: 1-59140-044-9; eISBN 1-59140-090-2, © 2003
- **Managing Multimedia-Enabled Technologies in Organizations**, S.R. Subramanya/ISBN: 1-59140-054-6; eISBN 1-59140-091-0, © 2003
- **Web-Powered Databases**, David Taniar and Johanna Wenny Rahayu/ISBN: 1-59140-035-X; eISBN 1-59140-092-9, © 2003
- **e-Commerce and Cultural Values**, Theerasak Thanasankit/ISBN: 1-59140-056-2; eISBN 1-59140-093-7, © 2003
- **Information Modeling for Internet Applications**, Patrick van Bommel/ISBN: 1-59140-050-3; eISBN 1-59140-094-5, © 2003
- **Data Mining: Opportunities and Challenges**, John Wang/ISBN: 1-59140-051-1; eISBN 1-59140-095-3, © 2003
- **Annals of Cases on Information Technology** – vol 5, Mehdi Khosrowpour/ ISBN: 1-59140-061-9; eISBN 1-59140-096-1, © 2003
- **Advanced Topics in Database Research** – vol 2, Keng Siau/ISBN: 1-59140-063-5; eISBN 1-59140-098-8, © 2003
- **Advanced Topics in End User Computing** – vol 2, Mo Adam Mahmood/ISBN: 1-59140-065-1; eISBN 1-59140-100-3, © 2003
- **Advanced Topics in Global Information Management** – vol 2, Felix Tan/ ISBN: 1-59140-064-3; eISBN 1-59140-101-1, © 2003
- **Advanced Topics in Information Resources Management** – vol 2, Mehdi Khosrowpour/ ISBN: 1-59140-062-7; eISBN 1-59140-099-6, © 2003

Excellent additions to your institution's library! Recommend these titles to your Librarian!

To receive a copy of the Idea Group Publishing catalog, please contact (toll free) 1/800-345-4332, fax 1/717-533-8661,or visit the IGP Online Bookstore at:
[http://www.idea-group.com]!
Note: All IGP books are also available as ebooks on netlibrary.com as well as other ebook sources. Contact Ms. Carrie Stull at [cstull@idea-group.com] to receive a complete list of sources where you can obtain ebook information or IGP titles.

The International Journal of Distance Education Technologies (JDET)

The International Source for Technological Advances in Distance Education

ISSN: 1539-3100
eISSN: 1539-3119

Subscription: Annual fee per volume (4 issues):
Individual US $85
Institutional US $185

Editors: Shi Kuo Chang
University of Pittsburgh, USA

Timothy K. Shih
Tamkang University, Taiwan

Mission

The International Journal of Distance Education Technologies **(JDET)** publishes original research articles of distance education four issues per year. JDET is a primary forum for researchers and practitioners to disseminate practical solutions to the automation of open and distance learning. The journal is targeted to academic researchers and engineers who work with distance learning programs and software systems, as well as general participants of distance education.

Coverage

Discussions of computational methods, algorithms, implemented prototype systems, and applications of open and distance learning are the focuses of this publication. Practical experiences and surveys of using distance learning systems are also welcome. Distance education technologies published in JDET will be divided into three categories, **Communication Technologies, Intelligent Technologies, and Educational Technologies**: New network infrastructures, real-time protocols, broadband and wireless communication tools, Quality-of Services issues, multimedia streaming technology, distributed systems, mobile systems, multimedia synchronization controls, intelligent tutoring, individualized distance learning, neural network or statistical approaches to behavior analysis, automatic FAQ reply methods, copyright protection and authentification mechanisms, practical and new learning models, automatic assessment methods, effective and efficient authoring systems, and other issues of distance education.

For subscription information, contact:

Idea Group Publishing
701 E Chocolate Avenue
Hershey PA 17033-1212, USA
cust@idea-group.com

For paper submission information:

Dr. Timothy Shih
Tamkang University, Taiwan
tshih@cs.tku.edu.tw

Just Released!

Designing Instruction for Technology-Enhanced Learning

Patricia Rogers
Bemidji State University, USA

When faced with the challenge of designing instruction for technology-enhanced education, many good teachers find great difficulty in connecting pedagogy with technology. While following instructional design practices can help, most teachers are either unfamiliar with the field or are unable to translate the formal design process for use in their own classroom. ***Designing Instruction for Technology Enhanced Learning*** is focused on the practical application of instructional design practices for teachers at all levels, and is intended to help the reader "walk through" designing instruction for e-learning.

The goal of ***Designing Instruction for Technology Enhanced Learning*** is to pool the expertise of many practitioners and instructional designers and to present that information in such a way that teachers will have useful and relevant references and guidance for using technology to enhance teaching and learning, rather than simply adding technology to prepared lectures. The chapters, taken together, make the connection between intended learning outcomes, teachings strategies, and instructional media.

ISBN 1-930708-28-9 (h/c) • US$74.95 • 286 pages • Copyright © 2002

"Most often, when forced to use new technologies in teaching, teachers will default to a technology-enhanced lecture method, rather than take advantage of the variety of media characteristics that expand the teaching and learning experience."
–*Patricia Rogers, Bemidji State University, USA*

It's Easy to Order! Order online at www.idea-group.com or call our toll-free hotline at 1-800-345-4332!
Mon-Fri 8:30 am-5:00 pm (est) or fax 24 hours a day 717/533-8661

Idea Group Publishing
Hershey • London • Melbourne • Singapore • Beijing

An excellent addition to your library